Molluscs

Edited by Genaro Diarte-Plata and Ruth Escamilla-Montes

Published in London, United Kingdom

IntechOpen

Supporting open minds since 2005

Molluscs
http://dx.doi.org/10.5772/intechopen.78418
Edited by Genaro Diarte-Plata and Ruth Escamilla-Montes

Contributors
Violeta Pardío-Sedas, Irma Wong, Leonardo Lizárraga, Karla López, Argel Flores, Guadalupe Barrera, Francisco Alarcón, Carlos Fernández, Igor A. Govorin, Simone Teixeira, Susmara Campos, Hanan Al-Khalaifah, Afaf Al-Nasser, Genaro Diarte-Plata, Ruth Escamilla-Montes, Salvador Granados-Alcantar

First published in London, United Kingdom, 2019 by IntechOpen
IntechOpen is the global imprint of INTECHOPEN LIMITED, registered in England and Wales, registration number: 11086078, The Shard, 25th floor, 32 London Bridge Street
London, SE19SG - United Kingdom
Printed in Croatia

British Library Cataloguing-in-Publication Data
A catalogue record for this book is available from the British Library

Additional hard copies can be obtained from orders@intechopen.com

Molluscs
Edited by Genaro Diarte-Plata and Ruth Escamilla-Montes
p. cm.
Print ISBN 978-1-83880-195-3
Online ISBN 978-1-83880-196-0
eBook (PDF) ISBN 978-1-83880-647-7

For EU product safety concerns:
IN TECH d.o.o., Prolaz Marije Krucifikse Kozulić 3, 51000 Rijeka, Croatia, info@intechopen.com or visit our website at intechopen.com.

Meet the editors

Genaro Diarte-Plata is a candidate level of SNI-CONACYT, Mexico and a biologist in aquaculture at the Universidad Autónoma de Sinaloa, Mexico. He holds a Master in Natural Resources and Environment from the Instituto Politécnico Nacional (IPN CIIDIR-Sinaloa), Mexico. He has been a professor at the IPN, Mexico, since 2006, and a researcher at the Sinaloa System of Researchers and Technologists (SSIT-INAPI Sinaloa). He also has a Doctorate in Marine and Coastal Sciences from the Universidad Autónoma de Baja California Sur, Mexico. He has conducted several investigations in the mariculture, culture, and ecology of molluscs and crustaceans. As a professor-researcher within the IPN, he has participated in 45 research projects and has given 10 seminars and 250 presentations at international and national conferences. He has published 5 book chapters and 17 indexed articles. He teaches descriptive and multivariate statistics at bachelor's level and aquaculture in postgraduate level. Dr. Diarte-Plata is also the director of 10 undergraduate theses and 3 master's and 1 doctorate from various universities.

Ruth Escamilla-Montes is a doctor in marine and coastal sciences with a specialty in aquaculture. She is also an assistant technologist at the Sinaloa System of Researchers and Technologists. Her areas of study are cultivation and ecology of mollusks and crustaceans. Over the last five years, she has focused on the isolation and characterization of lactic acid bacteria, yeasts, and bacilli with probiotic potential and pathogenic bacteria such as *Vibrio parahaemolyticus* and its effect on the growth, survival, and expression of genes of the immune system and stress of marine organisms. She has participated in 8 research projects, as well as presentations at national and international conferences, and has published 4 book chapters and 23 articles in ISI Thomson-JCR international journals. She is an arbitrator in the revision of manuscripts sent to international journals (ISI-JCR).

Contents

Preface

Mollusca phylum represents a great diversity of species in the animal kingdom. It is estimated that there are 117,358 species, which have a variety of sizes, shapes, and different lifecycles that allow them to inhabit different environments. The study of marine molluscs is attractive to paleontologists, zoologists, marine biologists and archeologists, collectors, and geneticists among other specialties. This is due to the vast complexity of types of characteristics that contribute to successful colonization in mangrove forests, coastal lagoons, reefs, and other areas, establishing ecological niches from the intertidal zone to slopes and ocean depths, including hydrothermal vents and trenches over 5000 m deep.

Bivalve molluscs have an important ecological importance in coastal ecosystems due to their filtering activity, since they help to improve the quality of the water and serve as food inside the trophic chain. Several species have economic interest for humans, in the fields of both food and ornamentation. For example, oysters have commercial importance for the formation of pearls: these appear when a grain of sand or a particle enters the shell of an oyster and the animal covers the grain with a substance called nacre.

This book consists of four sections. The first section "Introduction" offers information on mollusc generalities. In addition, these organisms are important in areas of commercial significance such as aquaculture and fishing. Similarly, molluscs have uses in pollution studies and environmental processes among others.

The second section deals with the "Social Aspects of Fisheries," which offers information on mollusc gathering in tropical regions of Brazil, where the role of women in fishing activities is recognized in terms of socioeconomic aspects as well as their work spaces. With regard to the relationships established between men and women, there is a need for economic production and social reproduction. Moreover, the scope of their traditional knowledge acquired and recognized on collecting shellfish and their perceptions regarding environmental changes involved in its activities are very important for planning environmental management strategies for estuarine areas and public policies to protect the environment.

The third section "Ecology" presents information on the predatory marine gastropod *Rapana venosa* (Valenciennes, 1846) in the northwestern Black Sea, including morphometric variations, imposex appearance, and biphallia phenomenon, and mentions that *R. venosa* presents a more elongated, thin, and less immense shell than other molluscs. The weight/size differences of *R. venosa* can confirm certain morphometric changes of the animal in different habitat conditions, as well as its evident morphometric characteristics and the appearance of impurities in females and the biphallia phenomenon in males. Marine waters subject to anthropogenic contamination make it possible to consider it as an indicator of the ecological status of marine coastal areas. Finally, the fourth section looks at aspects of the "Immune system" in two themes: the chapter "Immune Response of Molluscs" deals with the elements of the molluscan innate immune system and presents a case study of the immune response of the *Lymnaea stagnalis* mollusc against the parasite *Chaetogaster limnaei*. The effect of the parasite on some humoral immune response parameters such as nitric oxide, phenol oxidase, and lysozyme production is investigated. In conclusion, *L. stagnalis* exerts a humoral immune response against the *C. limnaei* parasite. However, this response is insufficient to eliminate the parasite. The chapter "Survival Differences of *Vibrio vulnificus* and *Vibrio parahaemolyticus* Strains in Shellstock Oysters (*Crassostrea virginica*) from Harvest to Sale: A Risk Perspective" provides information on estimating the risk associated with the consumption of raw oysters affected by contamination levels of *V. vulnificus* and *V. parahaemolyticus* and the temperature during postharvest and transport from Mandinga Lagoon to Mexico City. The results suggest that maintaining temperatures above 20°C during storage and transport of oysters increases the risk of infection by pathogenic strains.

The book is a document that can be consulted by students, professors, and researchers in areas related to the biological sciences.

Dr. Genaro Diarte-Plata and Dr. Ruth Escamilla-Montes
Instituto Politécnico Nacional (CIIDIR Sinaloa),
Mexico

Section 1

Introduction

Chapter 1

Introductory Chapter: Molluscs

Ruth Escamilla-Montes, Genaro Diarte-Plata and Salvador Granados-Alcantar

1. Introduction

Molluscs are one of the great animal phyla after arthropods. There are more than 90,000 living species and about 70,000 fossil species. It is a highly diverse group that includes eight classes: Caudofoveata, Solenogastres, Monoplacophora, Polyplacophora, Gastropoda, Bivalvia, Cephalopoda, and Scaphopoda, which have a wide variety of body shapes and structures [1]. They are quite simple molluscs, without eyes or shell, and even the most complex organisms on the planet. They are found in a great diversity of habitats, from the tropics to the polar seas and from the shallow muddy plains to the open ocean or the abyssal plains. They present diversity in their life strategies: they can be benthic, pelagic, or burial drills, molluscs originated in the sea, estuaries, fresh water and terrestrials [1–5].

Molluscs have a wide variety of food strategies, including herbivores, carnivores, predators, filter feeders, detritivores, and even some parasites [1], which are of great ecological importance due to their trophic relationships. They also have great ecological importance, since they are the link of pelagic and benthic processes, because they filter organic matter and phytoplankton from the water column, and their undigested remains, expelled as mucus or pseudofeces, become part of the sediment [6]. Several investigations showed strong evidence that natural populations of filtering bivalves can exert top-down control over phytoplankton in coastal areas [7].

The importance of molluscs as a fishing and aquaculture resource is well known, since there are enormously developed and productive fisheries of octopus, squid, cuttlefish, numerous species of bivalves, and some gastropods, as well as the use of various species in monitoring programs to analyze the effects of pollution and other disturbances on benthic communities [8, 9]. A large number of species have commercial importance since their meat is used as food and their shells as pieces of ornaments or in crafts. They are also useful as bioindicator of pollution or environmental processes and in the industry as a source of cosmetic and pharmaceutical products.

The mechanisms of defense in molluscs fulfill an important function against bacteria, fungi, protozoa, and metazoans. The immune response is mediated by cellular and humoral factors, and it has been shown that there is a close link between these two components. The hemocytes are cells capable of generating shape changes by the emission of pseudopodia, phagocytic activity, cytotoxicity, and encapsulation of large particles [10]. These circulating cells are responsible for generating different types of innate responses such as phagocytosis, encapsulation, production of cytotoxic substances, and antibacterial peptides. These cells are present in the hemolymph, but they are also able to leave the circulation and migrate to other tissues of the animal where they can be added to restrict infection or some tissue damage [11]. Its capacity of phagocytosis is one of the essential functions of

IntechOpen

hemocytes to eliminate exogenous agents such as bacteria or protozoa [12]; in this case, the production of reactive oxygen species is induced [13, 14]. In Pectinidae bivalves, the phagocytic capacity has been evaluated in *Pecten maximus* with various types of bacteria and yeasts [14].

The encapsulation allows the immobilization of larger particles than the hemocytes; this response involves the formation of concentric layers, formed by hemocytes. The formation of capsules of hemocytic origin against protozoan parasites has been studied in detail in the oyster *C. virginica* [15] and in the clams *Tapes semidecussatus* [16] and *Mercenaria mercenaria* [17]. The most studied model in the innate immune response of hemocytes against *the Schistosoma mansoni* trematode larvae has been the gastropod mollusc *Biomphalaria glabrata* (Mollusca: Pulmonata), where it has been shown that the parasite has a modulating effect on several hemocytic parameters, such as suppression of phagocytosis, change in mobility, variation in the number of hemocytes, cytoadherence, and metabolic capacity [18, 19].

Among the biggest challenges that arise in the culture of molluscs are the constant mortality events, which cause a significant reduction in production. The presence of diseases within cropping systems affects in particular the larval and post-larval stages in production laboratories, as well as juveniles and adults grown in the natural environment. Particularly in breeding places, the massive mortalities caused by diseases imply the total loss of production, with serious economic consequences. In most cases, studies have shown that the problems are caused by bacterial pathologies, being the main etiological agent members of gender *Vibrio* [20, 21]. In relation to the stages cultivated in natural banks, where the first studies focused on the pathologies are caused by parasitic protozoa, in recent years, research has focused on diseases of bacterial origin that affect the survival of crops.

At the present time, there is hardly any research being carried out on the bacterial populations that cause diseases which are associated with the culture of molluscs, and, therefore, there is little information on the subject, which has led to the search for alternatives aimed at the elimination of bacteria of crop water during the hatchery stages. Among the methods used in different water treatments as well as in chemotherapy, it has been observed that they are inadequate to avoid high mortalities. Therefore, the use of probiotic bacteria in mollusc culture is one of the most promising options in aquaculture, giving rise to a balanced bacterial population with self-regulatory capacity. In addition, their use avoids the dangers derived from antibiotics and other control measures, which help prevent diseases and avoid economic losses within this activity [22].

Molluscs can be used in monitoring plans, since these organisms have the peculiarity of having little or little movement, long life cycles, a high degree of tolerance to stress, an intimate relationship with sediment, and a rapid response to disturbances. Makes them ideal for the study of environmental changes of natural and anthropogenic origin.

Acknowledgements

The authors are grateful to Instituto Politécnico Nacional, Mexico (IPN CIIDIR Unidad Sinaloa), through of the Secretaría de Investigación y Posgrado, for the financial support (IPN SIP 20181467).

Conflict of interest

We declare no conflict of interest.

Author details

Ruth Escamilla-Montes[1], Genaro Diarte-Plata[1*] and Salvador Granados-Alcantar[2]

1 Instituto Politécnico Nacional, CIIDIR Sinaloa, Mexico

2 Universidad Autónoma de Occidente, Unidad Los Mochis, Sinaloa, Mexico

*Address all correspondence to: gdiarte@ipn.mx

References

[1] Hickman CP, Roberts LS, Keen SL, Larson A, L'Anson H, Eisenhour DJ. Integrated Principles of Zoology. 14th ed. New York, NY, USA: McGraw-Hill; 2011. 928 p

[2] Jackson JB. The ecology of *Thalassia testudinum*, Jamaica, West Indies. I. Distribution, environmental physiology and ecology of common shallow waters species. Bulletin of Marine Science. 1973;**23**:313-350

[3] Prieto AL, Ruiz J, García N. Diversidad malacológica de una comunidad de *Arca zebra* en Chacopata, Estado Sucre, Venezuela. Revista de Biología Tropical. 2001;**49**(2):591-598

[4] Reverol YM, Delgado JG, De Severeyn YG, Severeyn HJ. Embrionary and larval development of the marine clam *Tivela mactroides* (Bivalvia: Veneridae) from Caño Sagua Beach, Zulia State, Venezuela. Revista de Biología Tropical. 2004;**52**:903-909

[5] Marval J. Diversidad de moluscos en dos playas rocosas de la isla de Margarita [Tesis de Licenciatura en Biología]. Cumaná, Venezuela: Universidad de Oriente; 1986. 127 p

[6] Newell RI. Ecosystem influences of natural and cultivated populations of suspension-feeding bivalve molluscs: A review. Journal of Shellfish Research. 2004;**23**(1):51-62

[7] Dame RF. Ecology of Marine Bivalves: An Ecosystem Approach. 2nd ed. Boca Ratón, FL, EUA: CRC Press; 2011. 283 p

[8] Margalef R. Ecología. Ediciones Omega ed. Barcelona, España: S. A. Casanova; 1980. 951 p

[9] Guzmán Alvis A, Solano O. Estructura de la taxocenosis annelida-mollusca en la región de Mingueo, Guajira (Caribe Colombiano). Boletín de Investigaciones Marinas y Costeras. 1977;**26**:35-523

[10] Auffret M. Bivalve haemocyte morphology. In: Fischer WS, editor. Diseases Processes in Marine Bivalve Molluscs. Washington: American Fisheries Society; 1988. pp. 1690-1177. Special Publication N° 18

[11] Bachère E, Gueguen Y, Gonzalez M, de Lorge-ril J, Garnier J, Romestand B. Insights into the anti-microbial defense of marine invertebrates: The Penaeid shrimps and the oyster *Crassostrea gigas*. Immunology Reviews. 2004;**198**:149-168

[12] Bayne CJ. Phagocytosis and non-self-recognition in invertebrates. Phagocytosis appears to be an ancient line of defense. Bioscience. 1990;**40**:723-731

[13] Nappi AJ, Ottaviani E. Cytotoxicity and cytotoxic molecules in invertebrates. BioEssays. 2000;**22**(5):469-480

[14] Lambert C, Nicolas JL. Specific inhibition of chemiluminescent activity by pathogenic vibrios in hemocytes of two marine bivalves: *Pecten maximus* and *Crassostrea gigas*. Journal of Invertebrate Pathology. 1998;**71**(1):53-63

[15] Cheng T C. Hemocytes: Forms and functions. En:Kennedy VS, Newell RIE, Eble AF (eds.). The Eastern Oyster. College Park, MD: Maryland Sea Grant College, University of Maryland System, 1996, pp. 299-333

[16] Montes JF, del Rio JA, Durfort M, García-Valero J. The protozoan parasite *Perkinsus atlanticus* elicits a unique defensive response in the clam

Tapes semidecussatus. Parasitology. 1997;**114**:339-349

[17] Smolowitz R, Leavitt D, Perkins F. Observations of a protistan disease similar to QPX in *Mercenaria mercenaria* (Hard Clams) from the coast of Massachusetts. Journal of Invertebrate Pathology. 1998;**71**:9-25

[18] Noda S, Loker ES. Effects of infection with *Echinostoma paraensei* on the circulating haemocyte population of the host snail *Biomphalaria glabrata*. Parasitology. 1989;**98**:35-41

[19] Fryer SE, Bayne CJ. Opsonization of yeast by the plasma of *Biomphalaria glabrata* (Gastropoda): A strain-specific, time-dependent process. Parasite Immunology. 1989;**11**:269-278

[20] Beaz-Hidalgo R, Balboa S, Romalde JL, Figueras MJ. Diversity and pathogenicity of *Vibrio* species in cultured bivalve molluscs. Environmental Microbiology Reports. 2010;**2**:34-43

[21] Paillard C, Le Roux F, Borrego JJ. Bacterial disease in marine bivalves, a review of recent studies: Trends and evolution. Aquatic Living Resources. 2004;**17**:477-498

[22] Romalde JL. Héroes y Villanos: Bacterias asociadas al cultivo de moluscos. Revista AquaTIC. 2012;**37**:45-59

Section 2

Social Aspects of Fisheries

Chapter 2

Mollusc Gathering in Tropical Regions of Brazil

Simone Ferreira Teixeira and Susmara Silva Campos

Abstract

Gathering bivalve molluscs is an important part of extractive fishing activities in the northeastern region of Brazil and is performed mainly by women. This study addresses the invisibility of the activity despite the labor effort and income generation these women represent. Depending on the community, these fisherwomen either practice all steps of the activity or only some processes, such as preparing and selling the product, but are always involved in some part of the productive process. Despite participating in the generation of income, the work of these mollusc gatherers is considered invisible, without prestige and given little or no value when compared to other fishing activities, especially those exercised by men. Mollusc gathering may seem to be a non-complex practice, but requires a variety of traditional knowledge that is passed from one generation to the next. Such knowledge reflects the intimate understanding these workers have of productive processes and the environmental dynamics of coastal artisanal fishing. In the majority of traditional communities, the difficulties lead to the discouragement of this activity as work for future generations. Thus, there is a need for the recognition of the spaces of female mollusc gatherers, considering the relations between the need for economic production and social reproduction with the egalitarian representation of these workers in the entities of social representation of the class of fishers.

Keywords: fisherwomen, mollusc gatherers, traditional communities, traditional knowledge, gender issues, northeastern Brazil

1. Introduction

Artisanal fishing accounts for more than 90% of fishing jobs and the resulting catches correspond to more than half of the fishing production throughout the world [1]. In Brazil, artisanal fisheries in coastal waters account for 52.5% of all extractive fishing [2]. In 2007, northeastern Brazil was the second largest producing region of the country in terms of extractive marine fisheries, with 96.3% the result of artisanal fishing activities [3]. Among the total estimated production from extractive marine fisheries in the northeastern region in the same year, molluscs accounted for 5.8% [3], demonstrating the importance of this resource for the region.

Mollusc gathering in northeastern Brazil is performed mainly by women, whether participating in all steps of the activity or only in some processes, such as preparing the catch for sales and selling. Despite playing important roles in the productive and economic processes, women continue to be overlooked in official fishery statistics. In Ref. [4], women represent 24.35% of the General Fishery

Registry. However, this figure may not represent the effective participation of female labor in these activities, since not all fisherwomen are registered.

Gender division in terms of labor is an important factor to analyze in the lives of fisherwomen who participate in paid mollusc gathering activities and also perform housework. Although directly responsible for the maintenance and well-being of the family, women do not receive pay or recognition for their household chores. The work of women in artisanal fishing is also still considered invisible and seen as a marginal activity without prestige in terms of household income [5, 6], whereas the role of men is culturally linked to the productive sector. In this context, fisherwomen are unappreciated with regard to both housework and fishing activities.

Women in coastal communities in northeastern Brazil exercise an income-generating activity: the gathering of bivalve molluscs in estuarine and mangrove areas, the productive organization of which is directly related to the use, management and dynamics of natural marine resources [7]. Fishery resources in these areas are indispensible to the subsistence of traditional coastal communities. Bivalve molluscs are one of the important exploited resources with ecological and socioeconomic value and constitute either the main source of income of the families involved in this activity or a complementary source to other fishing activities [8].

In northeastern Brazil, the main marine bivalve species of economic importance in mollusc gathering are *Anomalocardia flexuosa* (Linnaeus, 1767) ("carib pointed venus" clam); *Phacoides pectinatus* (Gmelin, 1791) (thick lucine); *Ctena orbiculata* (Montagu, 1808) (dwarf tiger lucine); *Austromacoma constricta* (Bruguière, 1792) (constricted macoma); *Crassostrea mangle* Amaral & Simone, 2014 (mangrove oyster or bagpipe oyster); *Crassostrea brasiliana* (Lamarck, 1819) (diving oyster); *Iphigenia brasiliensis* (Lamarck, 1818) (giant false coquina); *Leukoma pectorina* (Lamarck, 1818) ("coroa de frade"); *Mytella charruana* (d'Orbigny, 1842) (charru mussel); *Mytella guyanensis* (Lamarck, 1819) (mangrove mussel); *Cyrtopleura costata* (Linnaeus, 1758) (angelwing clam); *Brachidontes exustus* (Linnaeus, 1758)

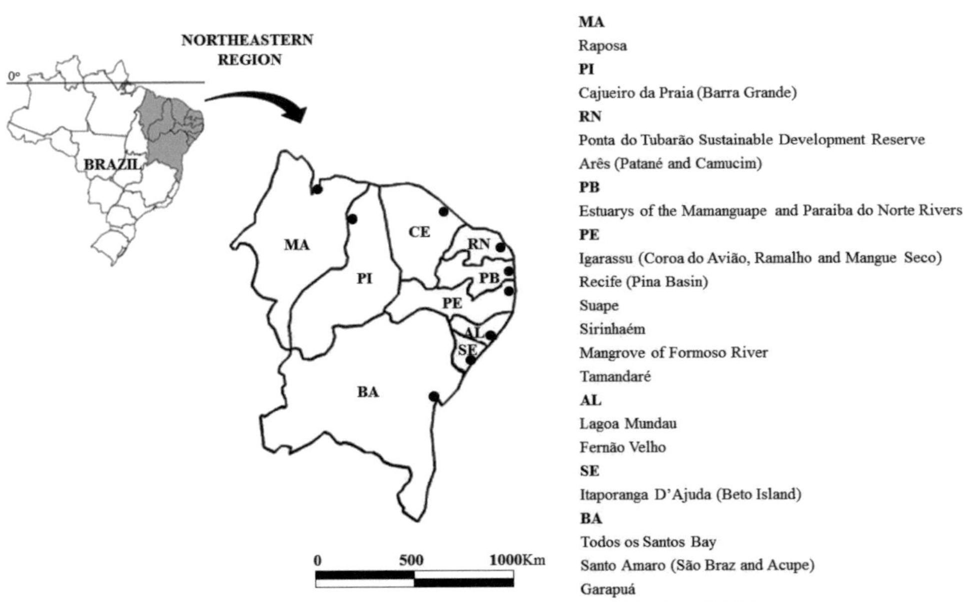

Figure 1.
Map showing sites where gathering of molluscs of commercial importance is performed in northeastern Brazil cited in present study. States: MA—Maranhão, PI—Piauí, RN—Rio Grande do Norte, PB—Paraíba, PE—Pernambuco, AL—Alagoas, SE—Sergipe and BA—Bahia. • State's capital.

(scorched mussel); and *Tagelus plebeius* (Lightfoot, 1786) (stout razor clam) [8–15, 17–20, 22–26, 28, 35, 38]. The species caught vary along the coast according to the type of substrate, which exerts an influence on the presence and abundance of species as well as catch processes.

The object of the present study was female bivalve mollusc gatherers on the northeastern coast of Brazil (**Figure 1**), particularly those in fishing communities of the Pina Basin in the city of Recife and the region of Suape in the city of Cabo de Santo Agostinho, both of which are located in the state of Pernambuco, where the authors have been developing research for more than 20 years, considering the work force and income generation of these women and, at the same time, their invisibility in extractive fishing activities.

2. Mollusc gathering in northeastern Brazil

In northeastern Brazil, bivalve mollusc gathering is the fishing activity that most employs women, but, nonetheless, these fisherwomen are both economically and socially marginalized, poor and have a low level of schooling, with little recognition among artisanal fishermen [8]. However, in the village of Acupe in the state of Bahia, the recognition of mollusc fisherwomen by other fishers has been occurring, probably due to the daily coexistence, as the mangrove is considered a place of interactions [16].

Brazilian legislation has made advances in terms of artisanal fishing in recent years. However, political policies continue to be out of step with the actual situation of mollusc gatherers, as evidenced by the lack of credit plans and accords, which impedes the purchasing of utensils and equipment for preparing the catch and aggregating value to the product.

Despite being the providers of many families, these working women of the tide go unseen by society. Moreover, mollusc gatherers face extreme working conditions, climate changes, the risk of accidents and illness due to the considerable physical effort exerted in unsanitary environments, which are reported as recurrent conditions in this activity, but mollusc gathering is an option where there is a lack of opportunity for formal labor. Moreover, the majority of gatherers also complement the family income with other types of work, since the activity does not ensure the sustenance of the family. These aspects serve to discourage this activity as work for future generations in the majority of traditional communities.

Another issue to consider is the participation of women who do not traditionally pertain to fishing communities, but have been gathering molluscs as a source of subsistence, especially in areas where there are large "pockets" of a marginalized population. This situation can lead to the marginalization of mollusc gathering and the possibility of its extinction as a traditional activity. Moreover, it can lead to overfishing due to the increase in effort and the lack of knowledge among these women regarding the proper techniques and processes of this activity.

In gender relations, which are passed down from generation to generation, there is the notion that fishing on the high seas is an activity that requires physical strength and is therefore considered men's work, whereas mollusc gathering is a lighter activity performed in mangroves and therefore considered women's work, as authors in Ref. [21] found among the residents of São Braz in the city of Santo Amaro in the state of Bahia. According to the author, the gender division in this community occurs due to the repulsion of the mangrove environment (mud), which is considered "dirty" and suitable for women's work, whereas the sea (water) is considered "clean" and a superior environment suitable for men's work. In the community of Beto Island in the municipality of Itaporanga D'Ajuda in the state of

Sergipe, the gender division is defined in work spaces, where men practice fishing and women gather molluscs and other products on the beach as well as in mangroves, lakes and rivers [22]. The same is found in the community of Cajueiro da Praia in the state of Piauí [23].

In northeastern Brazil, the predominance of women in mollusc gathering activities occurs in practically all communities, as seen in the municipality of Raposa in the state of Maranhão [24]; Cajueiro da Praia in the state of Piauí [23]; the Ponta do Tubarão Sustainable Development Reserve in the state of Rio Grande do Norte [11]; on the banks of the Paraíba do Norte River [25] and Mamanguape mangrove [26] in the state of Paraíba; in the Pina Basin in the city of Recife [12], in the mangroves of Formoso River [27], in the communities of Tamandaré and Sirinhaém [28], and in the region of Suape (reported by the authors) in the state of Pernambuco; in the community of Beto Island in the municipality of Itaporanga D'Ajuda in the state of Sergipe [22]; as well as in the mangroves of the Acupe district in the municipality of Santo Amaro [15, 16], in Todos os Santos Bay [29] and in the Canavieiras Extractivist Reserve [20] in the state of Bahia. In Barra Grande in the state of Piauí, mollusc gathering is a strictly female activity [17], as it is in the communities of Patané and Camucim in the municipality of Arês in the state Rio Grande do Norte, where these women are the wives of fishermen [18].

Therefore, mollusc gathering is an essentially female activity performed either individually or in groups composed of family members or neighbors who share transportation to the collection sites and the preparation of the catch for sale [8, 18]. One of the aspects that contribute to female participation in this activity is the possibility of working only a few hours per day close to home, which enables time for family and household tasks. In many communities, the groups are composed of the sons and daughters of the fisherwomen, which enables parental care, the transmission of knowledge on the techniques of the activity and an increase in the labor force.

The fishing community of Suape in the state of Pernambuco was the object of study by the authors in 1997 and 1998, when the area was considered one of the 12 main fishing colonies in the state [30, 31] and characterized as the most important coastal and estuarine zones in northeastern Brazil as a natural nursery for aquatic organisms that is practically pollution-free [32]. In 1979, work began on the Suape Port Industrial Complex, which required landfills and led to geomorphological and hydrodynamic changes [33]. The landfills of mangroves and possible pollution from the activities of the industrial complex exerted a direct influence on fishing production by compromising mollusc stocks as well as impeding the access of the fisherwomen to the molluscs. At the time, mollusc gathering was performed mainly by women and composed nearly completely by the shellfish *Anomalocardia flexuosa*, with rare catches of the giant false coquina (*Iphigenia brasiliensis*) and stout razor clam (*Tagelus plebeius*). Mollusc gathering was practiced by women who resided in the region as well as those from other municipalities, stimulated by the production of local stocks, which, however, were being affected by the impacts caused by the construction of the Suape Port. With the intensification of the enterprises at the Suape Port Industrial Complex, the socio-environmental impacts have deepened. The fisherwomen are being expelled from their territories and those who remain in the activity have seen their income diminished by the contamination of the molluscs and the decline in production [34]. Thus, the mollusc gatherers in the Suape region have been drastically affected and there is the real possibility of the extinction of this activity in one of the regions that was once considered to be among the most productive in the state of Pernambuco.

In the fishing community of Pina and Brasília Teimosa in the city of Recife, which has also been the object of study of the authors since 2006, the captures include the mussels *Mytella guyanensis* and *Mytella charruana*, the shellfish *Anomalocardia flexuosa*, *Phacoides pectinatus* and *Ctena orbiculata*, the constricted macoma (*Austromacoma constricta*), stout razor clam (*Tagelus plebeius*) and the oyster *Crassostrea* spp. This community is considered an urban fishing community, in which there is considerable real estate pressure due to its location being parallel to the coast and very close to the beach, which places pressure on fishing activities [8]. Moreover, this community can be considered a true "mangrove civilization," with its sociocultural and economic traditions linked to the estuary-mangrove environment [35].

Mollusc gathering may appear to be a non-elaborate practice with little planning and, therefore, less valued. However, this activity requires knowledge, meanings and interests constructed over time and recognized by the practices of the community, and the gatherers have a body of indispensible knowledge on the activity [8]. This knowledge is passed down from generation to generation, mainly by family members, and involves the planning and organization of actions, knowledge on species of mollusc, the use of different types of gear and the choice and location of collection sites [21, 29, 35]. Such knowledge reflects the intimate understanding these workers have of productive processes and the environmental dynamics of coastal artisanal fishing. Traditional knowledge on the activity is a product of artisanal fishing, as fishers generally have information on the environmental variables that exert an influence on the catch as well as hydrodynamics, seasonality, the management of fishing gear and the ecology of the target species, which is fundamental to successful fishing [16, 36, 37]. The artisanal fishermen and fisherwomen in the coastal area are also the keepers of knowledge on types of mangrove vegetation and the interactions of species with the mangrove [38]. These characteristics make the mollusc gatherers traditional peoples and communities.

In Brazil, traditional peoples and communities are protected by the National Sustainable Development Policy for Traditional Peoples and Communities, which was instituted in 2007 through Decree n° 6.040 issued on February 7, 2007 [39]. In this decree, traditional peoples and communities are considered to be "culturally differentiated groups, who recognize themselves as such, who have their own social organization and occupy and use territories and natural resources as a condition for their cultural, religious, ancestral and economic reproduction, using knowledge, innovations and practices generated and transmitted by tradition" [39].

Like all artisanal fishing, mollusc gathering involves passing traditional empirical knowledge from generation to generation. Such information acquired from traditional knowledge includes catch sites, adequate conditions for greater production and the preparation of the product for the market. The productive process involves gathering, washing, cooking, sorting, packaging, weighing, storage and commercialization, normally with variations depending on the species (**Figure 2**).

Seasonal and climate conditions are determinant factors for the traditional knowledge of mollusc gathering, since these aspects exert an influence on production and safe working conditions. In the Pina Basin in the city of Recife, the fisherwomen report that the best season of the year for mollusc gathering is the "dry period," referring to summer. According to the women, the rainy season leads to the death of molluscs in the areas due to the influx of freshwater from the rains and rivers that drain into the estuary. This report that the rainy season harms the mollusc gathering activity has been made by fisherwomen from the communities of Patané and Camucim in the municipality of Arês in the state of Rio Grande do Norte [18]. Besides fishing issues, the fisherwomen of the Pina Basin report that the rainy period is less favorable due to the risk to life caused by precarious working conditions in the rain.

The dynamics of the tides and lunar cycles also affect the catch. At low tide, the banks of sediment on which the molluscs are gathered (locally denominated "crôas" [sandbanks]) are exposed and enable the gathering activity, lasting, on average, 3–5 hours in the Pina Basin. In the region of Suape, gathering activities also occurred at

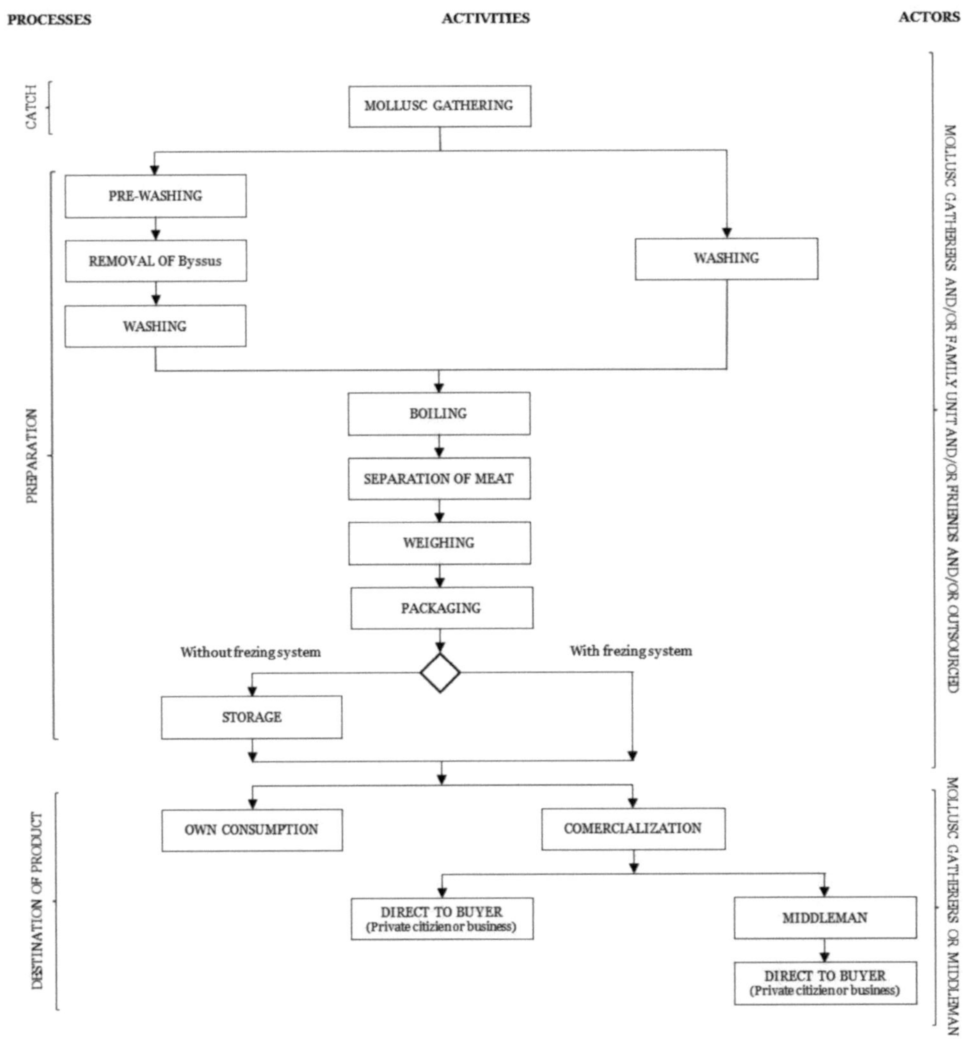

Figure 2.
Flowchart of processes, activities and actors involved in productive chain of gathering molluscs of commercial importance commonly practiced by fisherwomen in northeastern Brazil (Byssus: filamentous structure used to anchor mollusc to substrate).

low tide, when access to the banks of molluscs was possible, demonstrating knowledge of the tide variations to gain access to collection sites. The lunar phases recognized as the best for mollusc gathering are the full moon and new moon, which is in agreement with findings described in Ref. [15]. These correspond to the spring tides, which have a larger range of variation and are known locally as "the big tide."

There is also knowledge on the type of gear. The use of utensils varies based on the species targeted (**Table 1**) and type of substrate as well as the physical and age differences of the women. The fisherwomen normally use utensils that require less physical effort and cause less pressure on both the environment and the stocks, such as spoons, knives, cleavers, coconut shells, hoes, shovels, rakes, "bicheiros" (hooks formed by wooden pole with a curve iron rod at one end), hooks, sickles and spatulas or one's own hands, which, however, are less productive in comparison to "galeias" (empty plastic crates) (**Table 1**). In the community of Oiticica in the Canavieiras Extractivist Reserve in the state of Bahia, this activity was once practiced with the aid of cleavers, but began to be practiced with hoes in an effort

Sites in northeastern Brazil with mollusc fisherwomen	Bivalve mollusc species	Common name	Types of gear of fishing	References
Maranhão State				
Raposa	*Anomalocardia flexuosa*	Sarnambi	Cup; trawl nets; knives; rakes; hooks; spoons; trowel; one's own hands	[24]
	Iphigenia brasiliensis	Tarioba		
Piauí State				
Cajueiro da Praia (Barra Grande)	*Anomalocardia flexuosa*	Marisco	Shovels; spoons; one's own hands	[17, 23]
	Crassostrea mangle	Ostra	Knives; one's own hands	
	Iphigenia brasiliensis	Tarioba	One's own hands	
	Mytella charruana	Sururu; sururu-de-texto	Shovels; spoons; one's own hands	
	Mytella guyanensis	Sururu; sururu-de-dedo	Shovels; spoons; one's own hands	
	Tagelus plebeius	Tabaco-de senhora; pé-de-bode; bico-de-pato	One's own hands	
Rio Grande do Norte				
Ponta do Tubarão Sustainable Development Reserve	*Anomalocardia flexuosa*	Búzio; marisco	Spoons; buckets; knives; coconut shells; pans; rakes; one's own hands	[11]
	Mytella guyanensis	Sururu		
	Phacoides pectinatus	Búzio grande		
Arês (Patané and Camucim)	*Anomalocardia flexuosa*	Lilius	One's own hands; cleavers; "jereré or arrastão" (trawl nets with rakes)	[18]
	Crassostrea mangle	Ostra	No data	
	Mytella guyanensis	Sururu		
	Tagelus plebeius	Unha-de-velho		
Paraíba State				
Estuarys of the Mamanguape and Paraiba do Norte Rivers	*Anomalocardia flexuosa*	Marisco	No data	[13]
	Crassostrea brasiliana	Ostra de mergulho		
	Crassostrea mangle	Ostra de mangue; ostra gaiteira		
	Iphigenia brasiliensis	Taioba		
	Mytella charruana	Sururu de croa		
	Mytella guyanensis	Sururu		
	Tagelus plebeius	Unha-de-velho		
Estuary of the Paraíba do Norte River	*Anomalocardia flexuosa*	Marisco	Utensils; one's own hands	[25]
	Crassostrea mangle	Ostra de mangue		
	Mytella guyanensis	Sururu		
	Tagelus plebeius	Unha-de-velho		
Estuary of the Mamanguape River	*Anomalocardia flexuosa*	Marisco	One's own hands; spoons	[26]
	Crassostrea mangle	Ostra	Iron utensils; cleavers	
	Mytella guyanensis	Sururu	One's own hands; iron utensils; cleavers	

Sites in northeastern Brazil with mollusc fisherwomen	Bivalve mollusc species	Common name	Types of gear of fishing	References
Pernambuco State				
Igarassu (Coroa do Avião, Ramalho and Mangue Seco)	*Anomalocardia flexuosa*	Marisco	Net	[14]
Recife (Pina Basin)	*Anomalocardia flexuosa*	Marisco; berbigão; papa fumo; marisco de areia	"Galeia"; trowel; spoons; hoes; shovels; fork; one's own hands	Reported by the authors; [8, 12, 35]
	Austromacoma constricta	Marisco-casca-fina	"Galeia"; trowel; spoons; hoes; shovels; fork; one's own hands	
	Crassostrea mangle *Crassostrea brasiliana*	Ostra	Sickles; trowel; spatulas; knives; iron utensils	
	Ctena orbiculata	Marisco branco	"Galeia"; trowel; spoons; hoes; shovels; fork; one's own hands	
	Mytella charruana	Sururu de crôa; sururu; mexilhão do estuário	"Galeia"; trowel; hoes; one's own hands	
	Mytella guyanensis	Sururu; sururu raspado; mexilhão do estuário	Sickles; trowel; spatulas; hoes; knives; iron utensils	
	Phacoides pectinatus	Marisco-de-crôas; búzio grande; lambreta	"Galeia"; trowel; spoons; hoes; shovels; fork; one's own hands	
	Tagelus plebeius	Unha-de-velho	Hoes; "bicheiro"; hooks; iron utensils; knives; fork	
Suape	*Anomalocardia flexuosa*	Marisco	Hand nets or similar gear	Reported by the authors
	Iphigenia brasiliensis	Taioba		
	Tagelus plebeius	Unha-de-velho		
Sirinhaém and Tamandaré	*Anomalocardia flexuosa*	Mexilhão	Utensils; one's own hands	[28]
	Crassostrea mangle	Ostra		
Alagoas State				
Fernão Velho	No data	Marisco	No data	[40]
Sergipe State				
Itaporanga D'Ajuda (Beto Island)	*Anomalocardia flexuosa*	Maçunim	Spoons; knives; cleavers	[22]
	Mytella charruana	Sururu		
	Crassostrea brasiliana	Ostra	Cleavers	
Bahia State				
Todos os Santos Bay	No data	Ostra; sururu; lambreta; sarnambi; bebe-fumo, ralacoco; marisco; machadinho	Hoes; cleavers; "bicheiros"; one's own hands	[29]
Santo Amaro (São Braz)	No data	Ostra	Cleavers	[21]
		Marisco	Hooks	
		Sururu		

Sites in northeastern Brazil with mollusc fisherwomen	Bivalve mollusc species	Common name	Types of gear of fishing	References
Santo Amaro (Acupe)	*Anomalocardia flexuosa*	Marisco; bebe-fumo; papa-fumo	No data	[15, 38]
	Brachidontes exustus	Machadinha		
	Crassostrea mangle	Ostra; ostra de mangue; ostra de laje; ostra de mergulho		
	Cyrtopleura costata	Sururu-de-velho		
	Mytella charruana *Mytella guyanensis*	Sururu		
Garapuá	*Phacoides pectinatus*	Lambreta	No data	[19]
Canavieiras Extractivist Reserve	*Phacoides pectinatus*	Lambreta	Hoes	[20]

Table 1.
Bivalve mollusc species of commercial importance, common name, gear, sites in northeastern Brazil where molluscs are gathered and authors cited in present study.

to increase the catch rate, which has caused serious problems to the stocks [20]. In the community of the Pina Basin, male mollusc gatherers scrape the substrate with the edge of a plastic crate and then use the crate to sift and wash the molluscs. This process is more productive, but requires more physical effort and is therefore unviable for women. However, this method employed by the fishermen of the Pina Basin increases the pressure on the stocks, causing a decline in the species and impoverishing the area over time [8]. In the communities of Patané and Camucim, in the state of Rio Grande do Norte, fishing with the hands or machete was carried out with trawl net with rakes ("arrastão" or "jereré"), to increase harvesting, causing problems for reproduction of molluscs [18]. In the Suape community, mollusc production was conducted manually, with the aid of hand nets or similar gear.

The women are always involved in the preparation of the product for sale, working either alone or in groups that demonstrates the importance of women in the activity. Even in communities in which women do not participate in the collection process, they are responsible for the preparation and commercialization of the product. In the community of Brasília Teimosa, the presence of women occurs in both the capture and preparation stages of the molluscs [12]. The same occurred in the Suape region (studied by the authors) and on Beto Island in the municipality of Itaporanga D'Ajuda in the state of Sergipe [22], in the estuary of the Mamanguape River in the state of Paraíba [26], and in the community of Acupe in the state of Bahia [16]. In the community of Fernão Velho in the state of Alagoas, the preparation of molluscs and crustaceans as well as the salting of fishes are strictly performed by women [40]. Ref. [26] reports that the preparation for sales in Mamanguape occurs in the homes of the fisherwomen and, although generating income, it is considered an extension of housework, which further demonstrates the lack of recognition this work is given as a fishing activity.

The preparation of molluscs is quite rudimentary and normally involves boiling and separating the meat without adequate infrastructure under precarious hygiene-sanitation conditions, generally occurring around the residences of the fisherwomen. The boiling step can be considered quite unhealthy and taxing for these women, as it requires carrying firewood and inhaling a large amount of smoke. The separation

Sites in northeastern Brazil of mollusk gathering	Common name	Molluscs production	Sales price	References
Piauí State				
Cajueiro da Praia (Barra Grande)	Marisco	No data	R$ 7.00–10.00	[17]
	Sururu	315 kg/month (all community)	R$ 4.00–7.00	
Rio Grande do Norte				
Ponta do Tubarão Sustainable Development Reserve	Búzio	1–5 kg (mollusc meat/day/person) (mean: 3.2 kg)	R$ 2.50–6.00 (mean: R$ 3.53)	[11]
Arês (Patané and Camucim)	Lilius	5–20 kg (day/person; manual gathering) 5–8 sacks of 2–3 kg (fishing with net)	R$ 5.00/kg of mollusc meat (to middleman) R$ 7.00/kg of mollusc meat (to buyer)	[18]
	Ostra	No data	R$ 15.00–17.00/kg	
Paraíba State				
Estuaries of the Mamanguape and Paraiba do Norte Rivers	Ostra gaiteira	Extensively exploited	R$ 4.00–5.00/kg of mollusc meat	[13]
	Ostra de mergulho	Hard to gathering	R$ 15.00/100 units	
	Marisco	Extensively exploited	R$ 1.00–1.50/kg of mollusc meat	
	Sururu	Extensively exploited	R$ 1.50–2.00/kg of mollusc meat	
	Sururu de croa	Eventually exploited	R$ 1.00/kg of mollusc meat	
	Unha de velho	Extensively exploited	R$ 1.00–1.50/kg of mollusc meat	
Pernambuco State				
Igarassu (Coroa do Avião, Ramalho and Mangue Seco)	Marisco	230 sacks of 60 kg (all community)	R$ 2.00–5.00/kg of mollusc meat	[14]
Recife (Pina Basin)	Marisco	6.29 kg (mollusc meat/day/person; mean)	R$ 3.77/kg of mollusc meat (to middleman; mean) R$ 5.00/kg of mollusc meat (to buyer; mean)	[12]
	Sururu	7.38 kg (mollusc meat/day/person; mean)	R$ 4.37/kg of mollusc meat (to middleman; mean) R$ 5.06/kg of mollusc meat (to buyer; mean)	
	Ostra	2.22 kg (mollusc meat/day/person; mean)	R$ 4.43/kg of mollusc meat (to middleman; mean) R$ 7.35/kg of mollusc meat (to buyer; mean)	
	Unha de velho	4.90 kg (mollusc meat/day/person; mean)	R$ 4.12/kg of mollusc meat (to middleman; mean) R$ 4.71/kg of mollusc meat (to buyer; mean)	

DOI: http://dx.doi.org/10.5772/intechopen.83588

Sites in northeastern Brazil of mollusk gathering	Common name	Molluscs production	Sales price	References
Recife (Pina Basin)	Marisco	1–10 kg (mollusc meat/day/person)	US$ 1.56–4.38/kg of mollusc meat (to middleman) US$ 1.25–3.75/kg of mollusk meat (to buyer)	[35]
	Sururu	1–14 kg (mollusc meat/day/person)	US$ 1.56–3.75/kg of mollusc meat (to middleman) US$ 1.87–4.38/kg of mollusc meat (to buyer)	
	Unha de velho	1–5 kg (mollusc meat/day/person)	US$ 1.88–2.50/kg of mollusc meat (to middleman) US$ 2.50–5.00/kg of mollusc meat (to buyer)	
	Ostra	0.5–10 kg (mollusc meat/day/person)	US$ 1.88–6.25/kg of mollusc meat (to middleman) US$ 3.13–6.25/kg of mollusc meat (to buyer)	
Suape	Marisco	185.70–459 kg/month (all community) (mean: 316.2 kg*) *with shell	No data	Reported by the authors
Sirinhaém and Tamandaré	Marisco	2 kg (mollusc meat/person/day)	No data	[28]
Bahia State				
Todos os Santos Bay	Bebe-fumo	No data	R$ 4.00–10.00/kg of mollusc meat	[29]

R$: Real (BRL).

Table 2.
Production of mollusc species of commercial importance and sales prices, in mainly sites in northeastern Brazil where molluscs are gathered, and authors cited in present study.

of the meat is a slow, tiring process that can cause injuries to hands. Therefore, the preparation process is quite taxing for these women, as it involves physical effort in everything from the transporting the molluscs to the removal of the meat.

After preparation, the molluscs are packaged, weighed and sold either directly to the consumer or through middlemen (freelancers or representatives of fish markets), who may be individuals from the community itself or merchants from other locations. Fisherwomen with freezers have the option of freezing the molluscs with no need for immediate sale, as occurs with those who do not have storage conditions. This favors the direct sales of the product, since a lack of storage capacity requires selling to middlemen. Like the rest of the productive activity, commercialization is performed informally. A large part of the produce is sold, but it is also common for the fisherwomen and their families to consume the product.

The mollusc production and the sales price are showed in **Table 2**.

Besides selling the meat from the molluscs, the shells are used in many communities for craftwork. Among marine macrofauna, molluscs are often used in zoo-craftwork in the state of Pernambuco [41]. This activity has been consolidated as a new source of work and income and these craftspersons are found in the traditional craft markets in the city of Recife, Pernambuco [41]. In the community of Brasília Teimosa, the artisanal fishers have knowledge of the use of catches for the creation of zoo-craftwork, but report that they do not make these products themselves.

The raw material used for the creation of zoo-craftwork may pose the threat of further pressure on fish stocks. However, on the coast of the state of Paraíba [42] and in Brasília Teimosa [43], the market force that exerts pressure on fishing resources is food consumption. Thus, the zoo-craftwork in these places is made from catches used for culinary purposes or incidental catches and therefore does not place additional pressure on these resources [43].

Besides their traditional knowledge about their activity and the struggle for labor rights, the mollusc gatherers also report problems involving pollution and environmental degradation. Since they work in coastal areas, they are affected by the environmental problems that terrestrial water bodies have in Brazil. Regardless of the community, the fisherwomen list various impacts and environmental pressures, such as the destruction of mangrove areas, a lack of adequate organization and monitoring of the activity, trash dumping and the release of sewage, which affect the quality of local biodiversity and the river populations that survive on these resources [8]. Moreover, the installation of ports and large chemical and petrochemical plants in areas of considerable biological productivity affect fishing productivity, as observed by the authors in the 1990s in the area of Suape in the state of Pernambuco and in Mundau Lagoon in the state of Alagoas [44]. The impacts caused by the construction of the port in the area of Suape include the obstruction and/or extinction of natural nurseries for various species, the landfill of fishing grounds and areas of mangrove as well as a change in the type of sediment.

3. Conclusions

Women have been seeking room in fishing activities in terms of both socioeconomic aspects and the recognition of their work spaces. It is necessary to understand the relations that are established between men and women as well as relations between the need for economic production and social reproduction. As mollusc gathering is a predominantly female activity, the importance of gender in fishing activities must be recognized. However, the recognition and appreciation of these fisherwomen will only be possible when there is egalitarian representativeness of these laborers in the entities of social representation of the class of fishers, with their effective participation in fishing colonies, federations and confederations so that they can have the opportunity to discuss their needs and seek the rights and social protections inherent to their activities.

Moreover, the scope of their acquired and recognized traditional knowledge on mollusc gathering and their perceptions regarding the environmental changes involved in their activities are of extreme importance to the planning of environmental management strategies for estuarine areas as well as public policies to protect the environment and these fishing populations in an egalitarian manner.

Acknowledgements

The authors are grateful to the fisherwomen of the communities of the Pina Basin and Suape region in the state of Pernambuco for sharing the scope of their knowledge and their day-to-day living, always teaching us beyond what the literature has to offer.

Author details

Simone Ferreira Teixeira[1*] and Susmara Silva Campos[2]

1 University of Pernambuco, Recife, Pernambuco, Brazil

2 Federal Institute of Education, Science and Technology of Pernambuco, Recife, Pernambuco, Brazil

*Address all correspondence to: teixeirasf.upe@gmail.com

References

[1] FAO (Food and Agriculture Organization of the United Nations). Voluntary guidelines for securing sustainable small-scale fisheries in the context of food security and poverty eradication. Rome: FAO; 2015. 18 p. Available from: http://www.fao.org/3/a-i4356en.pdf [Accessed: 2018-10-08]

[2] Vasconcellos M, Diegues AC, Sales RR. Limites e possibilidades na gestão da pesca artesanal costeira. In: Costa AL, editor. Nas redes da pesca artesanal. Brasília: IBAMA; 2007. pp. 15-63

[3] IBAMA (Instituto Brasileiro do Meio Ambiente e dos Recursos Naturais Renováveis). Estatística da pesca 2007. Brasil: grandes regiões e unidades da federação. 2009. IBAMA: Brasília; 2009. 175 p. Available from: https://www.ibama.gov.br/sophia/cnia/livros/estatisticadepescadigital.pdf Accessed: 2018-10-08

[4] Fritsch, J. Memória do I Encontro Nacional das trabalhadoras da Pesca e Aqüicultura: Rumo à superação das desigualdades sociais. Brasil. In: SEAP-PR; 2004

[5] Cardoso DM. Mulher, Pesca e Ambiente. In: XXV Congresso Brasileiro de Ciências da Comunicação (INTERCOM); 1 a 5 Set 2002; Salvador/BA; 2002. Available from: http://www.portcom.intercom.org.br/pdfs/107751268139279009910777644676521911568.pdf [Accessed: 2018-10-08]

[6] Rosário JJ. Mulheres na pesca: espaço e cultura em Maragojipe/BA. Revista eletrônica Para Onde!? 2008. N.° 2. Available from: https://seer.ufrgs.br/paraonde/article/view/22070/12832 [Accessed: 2018-10-08]

[7] Oliveira NM. Rainha das águas, dona do mangue: um estudo do trabalho feminino no meio ambiente marinho. Revista Brasileira de Estudos de População. 1993;**10**(1/2):71-88. Available from: https://www.rebep.org.br/revista/article/view/493 [Accessed: 2018-10-08]

[8] Teixeira SF, Souza ACFF, Mariz D, Campos SS. Development pressures on urban and rural traditional fishing communities in Brazil: A case of tropical mollusk gatherers. In: Fletcher V, editor. Urban and Rural Developments: Perspectives, Strategies and Challenges. New York: Nova Science Publishers, Inc; 2016. pp. 39-66

[9] WoRMS Editorial Board. World Register of Marine Species [Internet]. 2018. Available from: http://www.marinespecies.org at VLIZ [Accessed: 2018-10-08]

[10] Amaral VS, Simone LR. Revision of genus Crassostrea (Bivalvia: Ostreidae) of Brazil. Journal of the Marine Biological Association of the United Kingdom. 2014;**94**(4):811-836. DOI: 10.1017/S0025315414000058

[11] Dias TLP, Rosa RS, Damasceno LCP. Aspectos socioeconômicos, percepção ambiental e perspectivas das mulheres marisqueiras da Reserva de Desenvolvimento Sustentável Ponta do Tubarão (Rio Grande do Norte, Brasil). Gaia Scientia. 2007;**1**(1):25-35. Available from: http://periodicos.ufpb.br/index.php/gaia/article/viewFile/2225/1953 [Accessed: 2018-10-08]

[12] Souza ACFF, Vieira DM, Teixeira SF. Trabalhadores da maré: conhecimento tradicional dos pescadores de moluscos na área urbana de Recife-PE (Tide workers: Traditional knowledge of shellfish fishermen in the urban area of Recife – PE). In: Alves RRN, Souto WMS, Mourão JS, editors. A Etnozoologia no Brasil: Importância, status atual e perspectiva. 1st ed. Vol. 4. Recife: NUPEEA; 2010. pp. 151-176

[13] Nishida AK, Nordi N, Alves RRN. Abordagem etnoecológica da coleta de moluscos no litoral paraibano (etnoecológica approach to collecting shellfish in Paraíba coast). Tropical Oceanography. 2004;**32**(1):53-68. Available from: https://periodicos.ufpe.br/revistas/TROPICALOCEANOGRAPHY/article/view/5034/4318 [Accessed: 2018-10-08]

[14] El-Deir SG. Estudo da mariscagem de Anomalocardia brasiliana (Mollusca: Bivalvia) nos bancos de Coroa do Avião, Ramalho e Mangue Seco (Igarassu-Pernambuco, Brasil) [thesis]. Recife: Federal University of Pernambuco; 2009. Available from: https://repositorio.ufpe.br/handle/123456789/8249 Accessed: 2018-10-17

[15] Souto FJB, Martins VS. Conhecimentos etnoecológicos na mariscagem de moluscos bivalves no manguezal do Distrito de Acupe, Santo Amaro–BA. Revista Biotemas. 2009;**22**(4):207-218. DOI: 10.5007/2175-7925.2009v22n4p207

[16] Souto FJB. A ciência que veio da lama: uma abordagem etnoecológica das relações ser humano/manguezal na comunidade pesqueira de Acupe, Santo Amaro-BA [thesis]. São Carlos: Universidade Federal de São Carlos; 2004. Available from: https://repositorio.ufscar.br/handle/ufscar/1755?show=full Accessed: 2018-10-08

[17] Freitas ST, Pamplin PAZ, Legat J, Fogaça FHS, Barros RFM. Conhecimento tradicional das marisqueiras de Barra Grande, área de proteção ambiental do delta do Rio Parnaíba, Piauí, Brasil. Ambiente & Sociedade. 2012;**15**(2):91-112. DOI: 10.1590/S1414-753X2012000200006

[18] Silva LKT. Pesca artesanal entre mudanças socioambientais: estudo de caso na APA Bonfim-Guaraíra/RN-Brasil [dissertation]. Natal: Universidade Federal do Rio Grande do Norte; 2015. Available from: https://repositorio.ufrn.br/jspui/handle/123456789/19857 Accessed: 2018-10-09

[19] Rondinelli SF. A exploração da lambreta, *Lucina pectinata* (Bivalvia, Mollusca), nos manguezais de Garapuá – Baixo Sul da Bahia, Brasil [dissertation]. Salvador: Universidade Federal da Bahia; 2009. Available from: http://www.dominiopublico.gov.br/pesquisa/DetalheObraForm.do?select_action=&co_obra=167123 Accessed: 2018-10-09

[20] Cavalcante AL, Pires MM, Strenzel GMR, Ferraz MIF. A arte da pesca: análise socioeconômica da Reserva Extrativista de Canavieiras, Bahia. Informe Gepec. 2013;**17**(2):81-99. Available from: http://e-revista.unioeste.br/index.php/gepec/article/view/5805/7269 [Accessed: 2018-10-09]

[21] Calazans RS. Marisqueira é pescadora: mulheres negras do Quilombo De São Braz – Santo Amaro, Bahia. Revista da Associação Brasileira de Pesquisadores/as Negros/as (ABPN). 2017;**9**(23):82-108. Available from: <http://www.abpnrevista.org.br/revista/index.php/revistaabpn1/article/view/508> [Accessed: 2018-10-09]

[22] Martins MLS, Alvim RG. Perspectivas do trabalho feminino na pesca artesanal: particularidades da comunidade Ilha do Beto, Sergipe, Brasil. Boletim do Museu Paraense Emílio Goeldi. Ciências Humanas. 2016;**11**(2):379-390. Available from: http://www.scielo.br/pdf/bgoeldi/v11n2/1981-8122-bgoeldi-11-2-0379.pdf [Accessed: 2018-10-09]

[23] Nascimento MSV, Sassi R. Análise da atividade pesqueira e das condições sócioeconomicas dos pescadores artesanais de Cajueiro da Praia, Estado do Piauí, Brasil. Gaia Scientia.

2007;**1**(2):141-154. Available from: http://www.periodicos.ufpb.br/index.php/gaia/article/view/2270/1996 [Accessed: 2018-10-09]

[24] Monteles JS, Castro TCS, Viana DCP, Conceição FS, França VL, Funo ICSA. Percepção socio-ambiental das marisqueiras no município de Raposa, Maranhão, Brasil. Revista Brasileira de Engenharia de Pesca. 2009;**4**(2):34-45. DOI: 10.18817/repesca.v4i2.141

[25] Nishida AK, Nordi N, Alves RRN. Aspectos socioeconômicos dos catadores de moluscos do litoral paraibano, Nordeste do Brasil. Revista de Biologia e Ciências da Terra. 2008;**8**(1):207-215. Available from: http://joaootavio.com.br/bioterra/workspace/uploads/artigos/22nishida-518174f45bd15.pdf [Accessed: 2018-10-09]

[26] Rocha MSP. Mulheres, manguezais e a pesca no estuário do Rio Mamanguape, Paraíba [dissertation]. João Pessoa: Universidade Federal da Paraíba; 2010. Available from: https://repositorio.ufpb.br/jspui/handle/tede/4586?locale=pt_BR Accessed: 2018-10-17

[27] Pedrosa BMJ, Lira L, Maia ALS. Pescadores urbanos da zona costeira do estado de Pernambuco, Brasil. Boletim do Instituto de Pesca. 2013;**39**(2):93-106. Available from: http://www.pesca.sp.gov.br/39_2_93-106.pdf [Accessed: 2018-10-17]

[28] Pedrosa BMJ. Pesca artesanal e áreas marinhas protegidas em Pernambuco: uma abordagem multidimensional e institucional [thesis]. Recife: Universidade Federal Rural de Pernambuco; 2016. Available from: http://www.tede2.ufrpe.br:8080/tede/bitstream/tede2/7075/2/Beatriz%20Mesquita%20Jardim%20Pedrosa.pdf Accessed: 2018-10-18

[29] Bandeira FPSF, Brito RRC. Comunidades pesqueiras na Baía de Todos os Santos: aspectos históricos e etnoecológicos. In: Caroso C, Tavares F, Pereira C, editors. Baía de Todos os Santos: aspectos humanos [online]. Salvador: EDUFBA; 2011. pp. 291-325. Available from: http://books.scielo.org/id/jy7mt/11 Accessed: 2018-10-17

[30] Motta R. O povoado de Suape: economia, sociedade e atitudes. 1978. 63 p

[31] IBAMA-NE/CEPENE. Boletim estatístico da pesca marítima no estado de Pernambuco no período de 1993 a 1994. 1995. 36 p

[32] CONDEPE (Instituto de Desenvolvimento de Pernambuco). Caracterização do complexo estuarino-lagunar da área de Suape (Pernambuco - Brasil). Síntese Ecológica. Recife; 1983. v. 1. 107 p

[33] Koening ML. Ecologia e dinâmica do fitoplâncton no estuário do Rio Ipojuca, após a implantação do porto de Suape (PE-Brasil) [thesis]. Recife: Universidade Federal Rural de Pernambuco; 1997

[34] Santana dos Santos MO, Alves SG, Silva JM, Mertens F, Gurgel IGD, Augusto LGS. Excluídas pelo desenvolvimento: mulheres e o Complexo Industrial Portuário de Suape. Revista de Geografia (Recife). 2016;**33**(3):117-140. Available from: file:///C:/Users/Simone/Downloads/229290-66495-1-PB.pdf [Accessed: 2018-10-09]

[35] Souza ACFF, Mariz D, Campos SS, Teixeira SF. Mollusk gatherers in a tropical urban estuary: Fishing activities and environmental perceptions. IOSR Journal of Humanities and Social Science. 2016;**21**(5):50-60. Available from: http://www.iosrjournals.org/iosr-jhss/papers/Vol.%2021%20Issue6/Version-1/G02106015060.pdf [Accessed: 2018-10-08]

[36] Teixeira SF, Mariz D, Souza ACFF, Campos SS. Effects of urbanization and the sustainability of marine artisanal fishing: A study on tropical fishing communities in Brazil. In: Ergen M, editor. Sustainable Urbanization. London: IntechOpen Limited; 2016. DOI: 10.5772/62785

[37] Ramalho CWN. Sentimento de corporação, cultura do trabalho e conhecimento patrimonial pesqueiro: expressões socioculturais da pesca artesanal. Revista de Ciências Sociais. 2012;**43**(1):8-27. Available from: http://www.periodicos.ufc.br/revcienso/article/view/417/399 Accessed: 2018-10-08

[38] Souto FJB. O bosque de mangues e a pesca artesanal no Distrito de Acupe (Santo Amaro, Bahia): uma abordagem etnoecológica. Acta Scientiarum. Biological Sciences. 2008;**30**(3):275-282. DOI: 10.4025/actascibiolsci.v30i3.5014

[39] BRASIL 2007. Decreto N° 6.040, de 7 de fevereiro de 2007. Institui a Política Nacional de Desenvolvimento Sustentável dos Povos e Comunidades Tradicionais. Available from: http://www.planalto.gov.br/ccivil_03/_ato2007-2010/2007/decreto/d6040.htm [Accessed: 2018-10-08]

[40] Santos EC, Sampaio CLS. A pesca artesanal na comunidade de Fernão Velho, Maceió (Alagoas, Brasil): de tradicional a marginal. Revista da Gestão Costeira Integrada. 2013;**13**(4):513-524. DOI: 10.5894/rgci428

[41] Souto AM, Silva MA, Melo Júnior M, Paranaguá MN, Pinto SL. Zooartesanato comercializado em Recife, Pernambuco, Brasil. Revista Brasileira de Zoociências. 2006;**8**(2):99-109. Available from: https://zoociencias.ufjf.emnuvens.com.br/zoociencias/article/view/105/95 [Accessed: 2018-10-09]

[42] Dias TLP, Alves RRN, Léo-Neto NA. Zooartesanato marinho da Paraíba. In: Alves RRN, Souto WMS, Mourão JS, editors. A etnozoologia no Brasil: importância, status atual e perspectivas. Recife: Nupeea; 2010. pp. 513-534

[43] Vieira DM. Ecologia, uso e classificação dos peixes pelos pescadores artesanais de Brasíllia Teimosa, Recife-PE [dissertation]. João Pessoa: Universidade Federal da Paraíba; 2011

[44] Diegues ACS. Povos e Mares: Leituras em Sócio-Antropologia Marítima. São Paulo: NUPAUB-USP; 1995. 269 p

Section 3

Ecology

Chapter 3

The Predatory Marine Gastropod *Rapana Venosa* (Valenciennes, 1846) in Northwestern Black Sea: Morphometric Variations, Imposex Appearance and Biphallia Phenomenon

Igor A. Govorin

Abstract

The presented chapter will show the results of long-term researches (2004–2013), concerning the study of variability of the size-mass relationships in mollusks *Rapana venosa* from the northwestern part of Black Sea (Odessa region, Zmiyiny Island, Danube Delta, Karkinitsky, and Tendra Gulfs) and near the eastern coast of Crimea (Sudak Gulf). The comparative evaluation has been made on the relationships between the total mass of the mollusks, mass of its soft body, and wide shells on the one hand and the size of animals (shell height, diameter, and thickness) in each study area on the other hand. Furthermore, the study of the appearance of the imposex in female mollusks (the small "penis" presence), "biphallia phenomenon" in male individual, and potential dependence of the occurrence of such anomalies on the ecological state of the marine coastal areas will be presented.

Keywords: Mollusca, Gastropoda, marine rapa whelk, *Rapana venosa* (Valenciennes), morphometric characteristics, size-mass relationships, imposex in female mollusks, biphallia phenomenon, northwestern Black Sea, Ukrainian shelf

1. Introduction

The Asian marine gastropod *Rapana venosa* (Valenciennes, 1846) {= *Rapana thomasiana* Crosse, 1861 [1]} was introduced in the Black Sea accidentally, probably from the Sea of Japan with the ship ballast waters. The first finding of this mollusk in Novorossiysk Bay (the Caucasian coast, eastern part of the sea) was recorded in 1946 [2]. Since 1959, the *Rapana* individuals were also discovered in the northwestern part of the Black Sea (NWBS) and in the summer of 1971, on the Odessa coast, at first in after-storm deposits. Live specimens were encountered here in 1974 [3]. Today, the results of fundamental researches on the ecology and functional morphology of *Rapana* in the Black Sea are widely presented in literature. However, most of these publications describe mollusks from different biotopes of the eastern part of the Black

Sea from Crimean coast [4] to Caucasian coast of Russia [5–8] or the shelf of Turkey [9, 10]. Thus, morphometric characteristics of *Rapana* in the NWBS concern the animals which inhabit near Romanian [11, 12] and Bulgarian coasts [13, 14].

The purpose of our work was to study the variability of the size-mass relationships in *Rapana venosa* from the Ukrainian shelf of the NWBS (Odessa region, Zmiyiny Island, Danube Delta), attributed to small depths and strong freshwater influence of the runoff of the large rivers—Dnieper, South Bug, Dniester, and Danube—in contrast to animals inhabiting near the eastern coast of Crimea (Sudak Gulf). Furthermore, the appearance of the imposex in female mollusks (presence of the small "penis") and "biphallia phenomenon" in male individual and tracking frequency of occurrence of these anomalies on ecological state of the marine coastal areas also were studied.

2. Material and methods

The living adult specimens of *Rapana venosa* were collected in 2004–2013 from the northwestern part of the Black Sea: near Odessa coast and neighboring areas, near Zmiyiny Island, Danube Delta area, Karkinitsky Gulf, Tendra Gulf, and from Ukrainian shelf of Crimea—in Sudak Gulf and near Tarkhankut Cape (eastern and western coast of Crimea) (**Figure 1**).

The animals were collected by diver on the rocks, stony bottom covered by mussels, on the sand, or on the sand with broken mussel shells. In the water area of the Odessa Sea Trade port, live specimens were collected on the surface of concrete breakwater constructions covered by mussel fouling. The empty *Rapana's* shells were found in after-storm deposits, mainly on the Odessa coast. A total of 529 live specimens of mollusks and 155 empty shells for the whole period of research were analyzed (**Table 1**).

The height (H, mm), diameter (D, mm) of shells, and thickness of its border (Th, mm) were measured in all individuals. The total mass of the animal (M_1, g), the raw mass of its soft body (M_2, g), and shell mass (M_3, g) were weighed. For

Figure 1.
The Black Sea map and sampling areas of Rapana venosa near the Ukrainian shelf of NWBS, Odessa region, Zmiyiny Island and Danube Delta, and western and eastern coast of Crimea (Tarkhankut Cape and Sudak Gulf).

Sampling site	Data	Number of specimens	Depth (m)	Substrate
Odessa coast and adjacent areas	2004–2013	284 (l.s.)	3.0–10.0	R, S, S&BMS, MF
		126 (e.s.)	—	—
Odessa Sea Trade port	18.07.2006	20 (l.s.)	5.0–7.0	CPC
Zmiyiny Island	2004–2008	140 (l.s.)	7.0–12.0	R, S, S&BMS, MF
Danube Delta	21.05.2005	1 (l.s.)	21.0	M
Karkinitcky Gulf	10.09.2008	5 (l.s.)	14.0	S, S&BMS
	17.06.2011	3 (l.s.)	6.0–10.0	S
	17.06.2011	29 (e.s.)	—	—
Tendra Gulf	07.07.2005	3 (l.s.)	4.0–5.0	S
Sudak Gulf, eastern Crimea	04.05.2004	60 (l.s.)	12.0–13.0	S
Tarkhankut Cape, western Crimea	20.09.2009	13 (l.s.)	10.0–11.0	S

Note: *Substrate: R, rocks; S, sand; S&BMS, sand with broken mollusks (mussels) shells; MF, mussel fouling on the bottom substrate; CPC, coast protecting concrete constructions (breakwaters and traverses); and M, bottom mud.*

Table 1.
The sampling sites and collected materials: live specimens (l.s.) of Rapana venosa and its empty shells (e.s.) from the after-storm ejects on the marine coast.

some mollusks of the Odessa coast, the dry mass of the soft body (M_4, g) was determined. The sex of the mollusks was determined by the presence or absence of the external genitals (penis): males, true females, and imposex females (with a small "penis"). Animal age (rounded up to a year) was calculated by the number of spawning marks on the shell [6].

The data of the morphometric parameters underwent standard statistical analysis (STATGRAPHICS Plus 5.0 program). Differences between the sampling sites in the measurements made were tested both by univariate and multivariate regression analysis of variance (ANOVA). On the basis of these data, statistically significant ($p < 0.01$) equations of mass-linear size of their shells from each of the study areas were drawn up.

3. Results and discussion

3.1 The morphometric features of *Rapana* in northwestern Black Sea (Ukrainian coast)

The height of shells (H) of the live specimens varied from 45.8 to 101.6 mm, its diameter (D) from 34.0 to 72.2 mm, and the thickness of the shell border (Th) from 0.9 to 4.2 mm. The total mass of the animal (M_1) was in the range of 20.6–173.9 g. The relation between the diameter, thickness of the shell, and the mollusk mass indices (total mass, raw and dry mass of the soft body, mass of shell), on the one hand, and the height of its shell, on the other hand, in each of the study areas of the Black Sea is described by the equation $y = a{\cdot}H^b$ (**Table 2**).

The mollusks from the NWBS were characterized by more elongated shells than specimens inhabiting the Crimean banks. The relation of shell height to its diameter (H/D) for *Rapana* near the Odessa coast, Zmiyiny Island, and in Sudak Bay was

Area	N	Min–max	ln a	b	r	R^2	SE
1	370	H = 45.80–101.60					
	370	D = 34.00–72.20	−0.6387	1.0661	0.986	97.30	0.051
	370	Th = 0.90–4.20	−3.9840	1.0609	0.835	69.76	0.202
	260	M_1 = 29.05–173.87	−8.9635	3.0582	0.990	98.09	0.128
	260	M_2 = 12.23–73.65	−13.0271	3.7951	0.979	95.96	0.234
	260	M_3 = 13.90–100.52	−8.5490	2.7869	0.987	97.58	0.150
	54	M_4 = 2.21–17.09	−14.0386	4.2661	0.910	94.02	0.156
2	60	H = 50.60–86.70					
	60	D = 36.50–69.30	−1.0625	1.1869	0.990	97.96	0.023
	60	Th = 1.40–4.80	−7.4876	1.9989	0.811	65.74	0.191
	60	M_1 = 20.65–133.08	−9.2083	3.1332	0.971	94.26	0.102
	60	M_2 = 4.05–37.89	−13.5531	3.8492	0.966	93.35	0.136
	60	M3 = 15.62 – 91.00	−8.6530	2.9078	0.948	89.82	0.129

Note: *This and in* ***Tables** 3 and* ***4****: N, number of specimens; H, D, Th, height, diameter of shell, and thickness of its border (mm); M_1, total biomass of the mollusk; M_2 and M_4, raw and dry mass of the soft body; M_3, shell mass (g); Min–max, the range of variability of the linear and mass indices; a, b, coefficients of the equations; r, coefficient of correlation; R^2, coefficient of determination (%); and SE, standard error of estimation.*

Table 2.
The parameters of the equations ($y = a{\cdot}H^b$) of the relation between the diameter (D), thickness of the shell border (Th), and the main indices of the mollusk mass (M_1–M_4), on the one hand, and height of its shell (H), on the other hand, in the Rapana venosa from Odessa region of the northwestern Black Sea (1) and southeastern Crimean coast, Sudak Gulf (2).

equal to 1.7 ± 0.02, 1.4 ± 0.02, and 1.3 ± 0.01, correspondingly. Similar differences were observed in the thickness of shell walls. In the Odessa area, the mean values in the relation of the height of the animals to their shell thickness (H/Th) were considerably higher than in those near the southeastern Crimean coast, 42.81 ± 0.64 and 26.02 ± 0.77, correspondingly. However, the smallest thickness of *Rapana* shells was recorded near the Danube Delta [15]. Mollusk shell H was 82.7 mm, D = 53.2 mm, and Th was only 1.3 mm (H/Th = 63.6).

Typical shells of similar size mollusks (H = 80.1–82.7 mm) from the study areas in the NWBS and Crimean coast are shown in **Figure 2**. The relations of H/D and H/Th for these specimens of *Rapana* are from the near Danube Delta area (A) = 1.55 and 63.61, Odessa coast (B) = 1.52 and 45.22, and Sudak Gulf, eastern Crimea (C) = 1.23 and 19.38, correspondingly.

The comparison of relations of the thickness of shell border (Th, mm), diameter of shell (D, mm), and shell mass (M_3, g) on the one hand to shell height (H, mm) on the other hand has shown the existing differences between the mollusks in the study areas. The shells of the Crimean *Rapana* specimens are significantly thicker and heavier than NWBS mollusks (**Figure 3**). As a consequence, the relation of the total mass to shell mass (M_1/M_3) was lower for the Crimean mollusks than for *Rapana* in the NWBS, 1.5 ± 0.01 and 1.9 ± 0.01, correspondingly. The relation of total mollusk mass to its raw body mass (M_1/M_2) was higher for *Rapana* from the Sudak Bay (3.6 ± 0.04) than for the animals from the NWBS (2.5 ± 0.03).

The shell mass (M_3) of *Rapana* near the Odessa coast and Zmiyiny Island was considerably lower than in specimens from Sudak, 51.5 ± 0.5, 52.6 ± 0.4, and 66.6 ± 0.5%, respectively, of total mass (M_1) of the animal. However, a lower shell mass was observed in the near Danube Delta area—43.1% of the total animal mass

Figure 2.
General view of the shells of Rapana venosa (Valenciennes, 1846) {= Rapana thomasiana Crosse, 1861 [1]} from different areas in the Black Sea: near Danube Delta (A), Odessa coast (B), and Sudak Gulf, eastern Crimea (C).

(M_1 = 79.8 g). It should be noted that for mollusks in the Kerch Strait (northeastern Crimean coast), this index was about 60% [16].

If we compare the diagrams of the dependence of the mollusk's soft body mass (M_2) on its total mass with shell (M_1) in NWBS and near the eastern coast of Crimea, a significant difference in the fatness of *Rapana* from these regions of the Black Sea can be clearly seen. So, the M_2 index was highest for the animals near the Zmiyiny Island and lowest for the mollusks in Sudak Gulf. The rapa whelk specimens near the Odessa coast occupied an "intermediate" position (**Figure 4A**).

At the same time, the proportion of shell mass in the total mass of the animal (M_3/M_1) was, on the contrary, the highest near the Crimean coast. Thus, it is necessary to pay attention to the almost complete coincidence of the trend lines in the graphs of the M_3/M_1 dependence for mollusks from Odessa and Danube regions of the Black Sea (**Figure 4B**).

In our opinion, a much thinner and less massive of *Rapana* shell in NWBS mollusks and near the Odessa coast, in particular, is ascribed to a low salinity in these areas. The mean of monthly salinities of coastal waters near Odessa, Sudak and Novorossiysk (place of origin of *Rapana* in the Black Sea) varies in the range of 11.70–15.73, 18.27–18.50, and 17.61–18.00 g/kg, accordingly [17]. In particular, near the Odessa coast, the average salinity of sea water in 1980–2009 was 13.8 g/kg, and in some abnormal years (2010), the average annual values of salinity could be reduced to 11.2 g/kg (in summer period 9.8 g/kg). It may suggest that the insufficient mineralization of sea waters stipulates a certain lack of building components for the mollusk shell.

At the same time, there are other opinions on the matter, excluding the direct dependence of the thickness and weight of the shells of mollusks on the level of salinity of the marine environment in the area of their habitat. According to I.P. Bondarev, comparative analysis of the thickness of the shells of *Rapana* from areas with different salinity (Azov-Black Sea basin) shows that the direct relationship between these parameters is absent [18]. This author believes that many factors may affect skeleton sizes in marine ectoderms, including in predation pressure, resource availability, energy acquisition, and the effort required to extract calcium carbonate ($CaCO_3$) from seawater. The main factor influencing the formation of thick-walled shell of the *Rapana venosa* from the Black Sea is the amplitude of the temperature fluctuations in the locality.

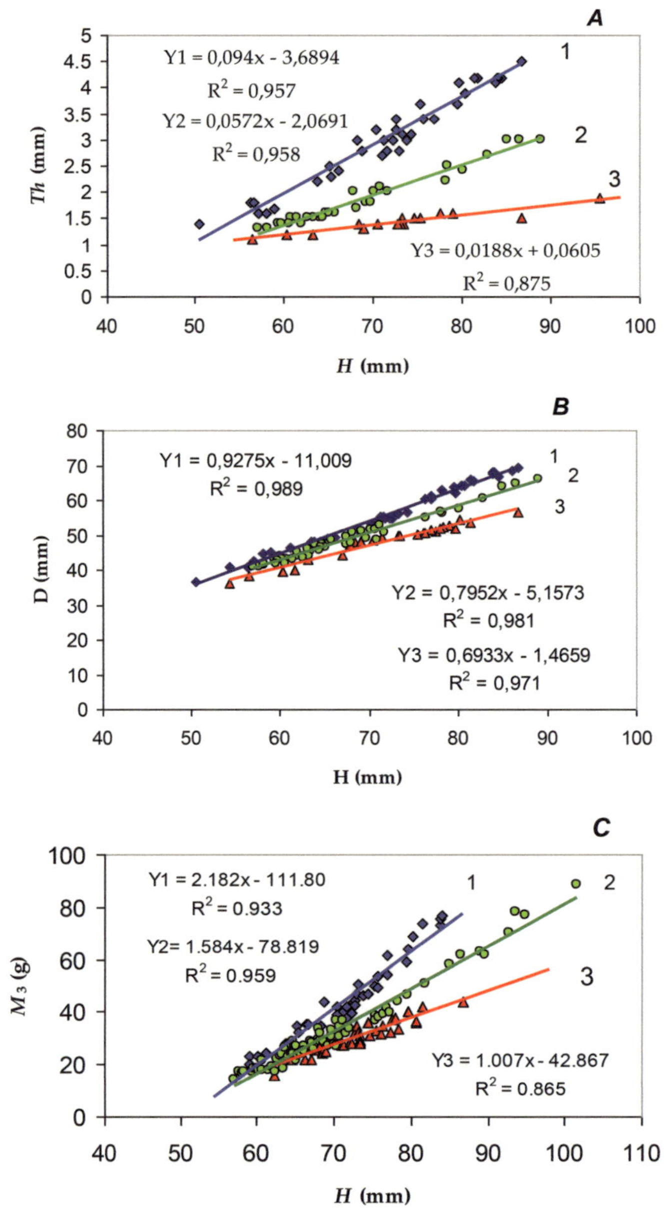

Figure 3.
The graphs of the relations of the thickness of the shell border (Th, mm) (A), diameter of shell (D, mm) (B), and the shell mass (M_3, g) (C) to the height of shells (H, mm) in Rapana venosa from Sudak Gulf, eastern Crimea (–◆– 1), near Zmiyiny Island, Danube region (–●– 2), and near Odessa coast (–▲– 3).

There is also an opinion that the ecological conditions in the different biotopes defined, first of all, the size and thickness of *Rapana* shells. The main ecological factors determining the morphological characteristics of the rapa whelk in these biotopes are their specific diversity and the abundance and size of their potential prey, in the NWBS—mainly bivalve mollusks *Mytilus galloprovincialis*, *Anadara* sp., and *Chamellia gallina* [8].

At the same time, in our opinion, in the northwestern part of the Black Sea, the food base of *Rapana* is not the main limiting factor, determining the mass of the mollusk's shells and its morphometric features, including the thickness of its walls. Thus, in the Odessa region of the NWBS, the main food component for *Rapana* is

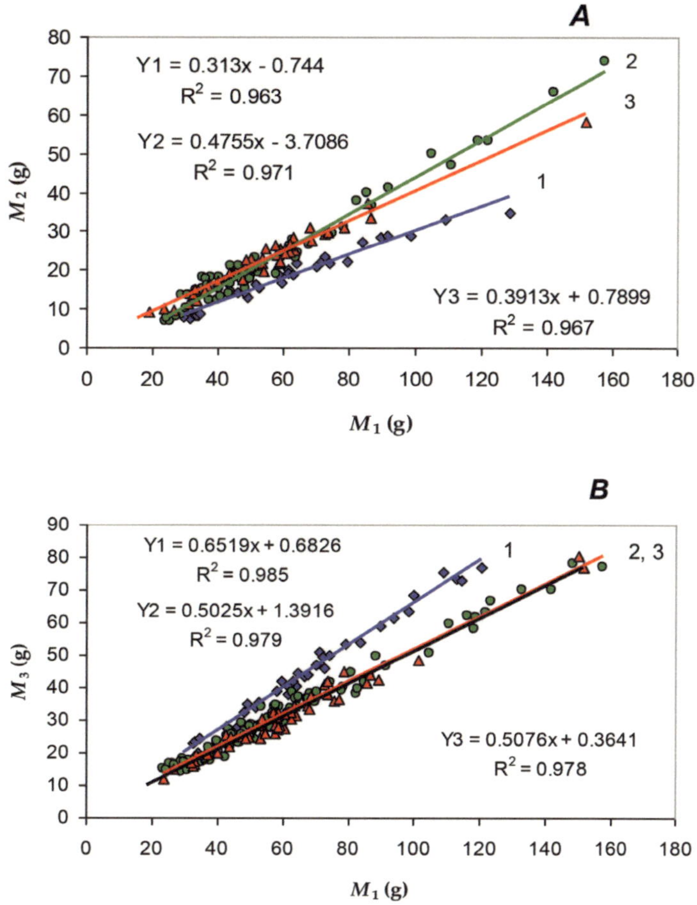

Figure 4.
The graphs of the relations of the raw mass of the soft body (M_2, g) (A) and the shell mass (M_3, g) (B) to the total mass of the animal (M_1, g) in Rapana venosa from Sudak gulf, eastern Crimea (–♦– 1), near Zmiyiny Island, Danube region (–●– 2), and near Odessa coast (–▲– 3).

mussel *Mytilus galloprovincialis* (Lamarck, 1819). According to our observations, the density of this bivalve mollusk in benthic settlements (depth 3 m) near the Odessa coast can reach from 2.325 ± 0.417 to 2.725 ± 0.394 thousand spec./m^2 in the spring and summer and its biomass from 15.93 ± 2.81 to 16.56 ± 1.98 kg/m^2. At the same time, almost one third of the mussels in the settlements (32.7 ± 6.5%) were animals of the size group 30–40 mm.

The study of the consumption rates of rapa whelk in natural environment showed that this predatory mollusk most actively consumes the mussels from 30 to 40 mm length (40.2 ± 3.2%). On average, one predator with a shell height (*H*) 60–80 mm eats for a day one mussel 36.3 ± 1.5 mm long with a total mass 3.94 ± 0.56 g (1.48 ± 0.22 g raw mussel meat) [19]. Therefore, the *Rapana* in this region of the NWBS does not experience difficulties with obtaining food, which is confirmed by the high levels of the fatness of mollusks in comparison with the Crimean specimens in Sudak Gulf.

The main function of mollusks' shell is protection of soft organs of animal from outer biotic and abiotic subjections, including the external pressure increasing with depth. In this regard, it should be noted that the relatively small depths in the NWBS coastal areas (11–15 m near Odessa coast and 10–25 m in the Danube area than depths of more than 30–50 m near the Crimean and Caucasian coasts) are favorable for the survival of *Rapana* specimens with thin and more fragile shells. It should be noted that in the 1970s, some researchers even proposed to refer the *Rapana* mollusks from the Odessa Bay to a separate ecological form, adapted to the local conditions differing from those of the eastern Black Sea [3].

The discovery in May 21, 2005, for the first time, a live adult male specimen of *Rapana* in the Danube Delta region can serve as confirmation of the further expansion of this mollusk in the NWBS. The mollusk was found 10 km from the mouth of the river (45°24.021′N; 29°53.022′E) at a depth of 21 meters on muddy bottom [15]. Prior to this case, there were no reports in the literature of the findings of *Rapana* in this water area experiencing a constant freshening effect of river flow. The extreme western point of the Odessa coast of NWBS, where we found in after-storm ejects the empty shells of this mollusk, is located on the sandy spit which separates the Tuzlovsky group of saltwater estuaries (Shagany, Alibey, Burnas) from the open sea (45°48.152′N; 30°06.154′E).

In comparison with the same size individuals from the southeastern coast of Crimea (Sudak) and the region of Zmiyiny Island, this specimen was more elongated. The ratio of the height of the shell to its diameter (H/D) in the animal found is 1.55. For comparison, in mollusks from the region of Zmiyiny Island and near the Crimean coast, this index is much lower (1.35 and 1.24, respectively). The shell of *Rapana* in the Danube region was characterized by a thinner wall (Th = 1.3 mm) and very less mass (M_3 = 34.4 g) (**Table 3**).

The final equations of the multiple regression analysis linking the basic mass parameters (M_1–M_3) and linear size of mollusks (H and D) for *Rapana venosa* in the Ukrainian area of the NWBS are

$$\log M_1 = -7.763 + 1.0724 \cdot \log H + 1.8556 \cdot \log D \quad (n = 240; R^2 = 98.91; SE = 0.0938) \tag{1}$$

$$\log M_2 = -2.651 + 1.159 \cdot \log M_1 + 0.2551 \cdot \log H - 0.0071 \cdot \log D \quad (n = 206; R^2 = 96.99; SE = 0.1969) \tag{2}$$

$$\log M_3 = -0.973 + 0.7918 \cdot \log M_1 - 0.2541 \cdot \log H + 0.5792 \cdot \log D \quad (n = 212; R^2 = 98.47; SE = 0.1026) \tag{3}$$

Since the p-value in the ANOVA table is less than 0.01, these mathematical models show the statistically significant relationship between the variables at the 99% confidence level and can be used for determining the mollusk indices of the soft body and of the shell from this area without dissection of the animals. In addition, these equations can be used to predict the vital mass of mollusks (M_1) based on the size-mass characteristics of empty shells from after-storm ejects on the coast [20].

3.2 The imposex appearance in female specimens of *Rapana* from different marine coast areas on the Ukrainian shelf of Black Sea

The *Rapana venosa* is a bisexual marine mollusk. The males can be easily differentiated from females by outward copulative organ (penis) that is used for "capture" and holding of partners during mating. Its length is about 15–20 mm in the adult male. Sexual affinity can be visually determined by the height of the mollusk shell that is 30 mm higher for males. Females of *Rapana* with signs of masculinization (a phenomenon of "imposex") that inhabit the muddy marine areas are an exception. Imposex females are characterized by the development of male characteristics, such as structure that looks like a small "penis."

Many researchers have the opinion that this abnormality is caused by the exposure of those animals to organotin compounds (Ots), mainly tributylin (TBT) and

Area (data)	N	H	D	Th	M_1	M_2	M_3
1 (21.05.2005)	1	82.7	53.2	1.3	79.8	35.2	34.4
2 (06.08.2004)	4	81.0 ± 0.4	53.3 ± 0.5	1.9 ± 0.1	78.5 ± 0.9	36.1 ± 3.2	39.3 ± 3.3
3 (06.08.2004)	4	82.3 ± 0.9	60.8 ± 1.1	2.4 ± 0.1	104.9 ± 6.1	47.7 ± 2.6	52.4 ± 3.6
4 (04.05.2004)	7	82.9 ± 0.6	66.6 ± 0.5	3.9 ± 0.1	105.5 ± 4.1	32.3 ± 1.1	67.5 ± 3.4

Table 3.
The size-mass characteristics of Rapana venosa from Danube Delta region of the northwestern Black Sea (1), near the Odessa coast (2), Zmiyiny Island (3), and in Sudak Gulf, eastern coast of Crimea (4), 2004–2005.

triphenyltin (TPT), of the antifouling paints, used worldwide to cover the boats and other metallic constructions in order to prevent corrosion [21–23]. Owing to the environmentally deleterious properties of Ots, TBT-based antifouling paints were banned by the International Maritime Organization in 2008; however, these paints are still widely used [24]. In the last years, imposexed females of *Rapana* were found in the Northern Adriatic Sea [25] and on the Atlantic Ocean coast, Chesapeake Bay, USA [26]. Generally, imposex has been previously reported in more than 30 caenogastropods in South America [27].

In the Black Sea, this phenomenon has been discovered for the first time near Odessa coast, NWBS region [28]. Among the 135 specimens of female *Rapana* from NWBS, examined in 2004–2010 for the presence of the phenomenon of imposex ("masculinization" of females), we found 17 anomalous females (12.6%) with a small rudimentary "penis" no longer than 4–5 mm in length. All mollusks with signs of imposex were adult individuals, the age of which varied from 3 to 7 years.

The lowest occurrence of imposex among female *Rapana* was observed in the region of Zmiyiny Island—1.5% (1 specimen from 65 females examined). Near the Odessa coast, this index was 6.5% (7 imposex specimens from 61), and in the water area of the Odessa Trade Port, 100% of the females (9 specimens) had an underdeveloped small "penis." As the manifestation of intersexuality in other species of gastropods of the Black Sea was most often found in polluted marine bays [4], the prevalence of masculinized females in Odessa port becomes clear.

According to their morphological characteristics, the imposex females distanced from males and typically female individuals. Thus, the total mass of the individual (M_1) and the mass of the shell (M_3) in the anomalous females were higher than in the same size animals of both sexes, and on the wet and dry mass of its soft body (M_2 and M_4), they occupied an intermediate position between these groups of mollusks (**Table 4**).

The results we obtained about the morphometric differences in abnormal females of *Rapana* support the opinion that imposex individuals represent a kind of "intermediate" between typical females and males, since the changes occurring with them can affect not only the external sexual characteristics of the animal (the presence of a rudimentary "penis" in particular) but also the structure of its internal organs [4].

The clear differences in the graphs of the relations between the dry mass of a soft body of mollusks (M_4, g) and the size of its shell (H, mm) in males and typical females of rapa whelks on one hand and females with signs of imposex on the other hand near the Odessa coast can serve as a visual confirmation of this opinion (**Figure 5**).

As seen in the figure, the trend line of the depending M_4/H in "abnormal" females with signs of masculinization (fm) occupies an intermediate position, distancing themselves from these dependency graphs for males (m) and typical females (f). Unfortunately, a small number of abnormal females (only 17

Sex	N	Lim H	Lim M	ln a	b	R^2	SE
Male	21	66.7–81.4	M_1 = 37.22–92.75	−9.567	3.150	91.48	0.106
			M_2 = 15.81–38.22	−17.322	4.795	97.60	0.134
			M_3 = 19.60–49.67	−9.263	2.969	65.83	0.134
			M_4 = 4.96–9.22	−7.595	2.193	68.57	0.154
Female	12	62.2–74.6	M_1 = 32.25–58.47	−8.517	2.913	88.13	0.085
			M_2 = 13.60–23.85	−8.854	2.775	62.76	0.141
			M_3 = 16.92–32.58	−7.137	2.452	61.94	0.127
			M_4 = 5.36–9.47	−9.355	2.635	64.25	0.141
Imposex female	12	63.2–84.2	M_1 = 33.49–98.16	−10.820	3.465	92.50	0.095
			M_2 = 12.18–40.11	−12.206	3.575	86.40	0.122
			M_3 = 19.70–53.42	−9.616	3.067	91.92	0.097
			M_4 = 3.26–12.83	−15.023	3.937	87.04	0.161

Table 4.
The parameters of the equations ($M = a \cdot H^b$) of the relation between the mass indices of mollusks (M_1–M_4), on the one hand, and height of its shell (H), on the other hand, in the same size male, female, and imposex female species of Rapana venosa in Odessa region of the northwestern Black Sea, July–August 2004–2009.

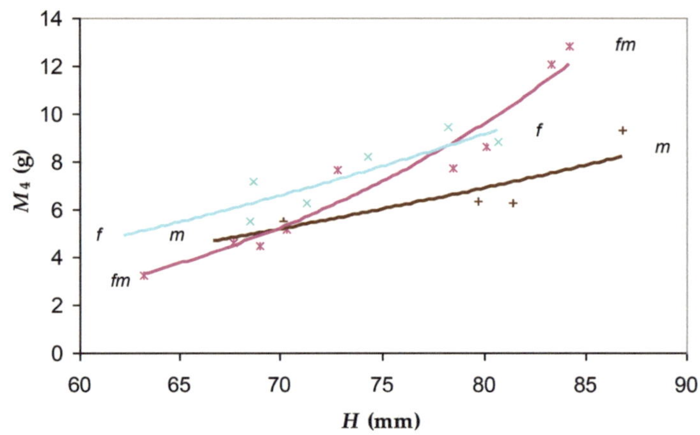

Figure 5.
The graphs of the relations between the dry mass of the soft body of mollusks (M_4, g) and its shell height (H, mm) in Rapana venosa near the Odessa coast: m, male specimens (+); f, typically female individuals (×); and fm, imposex females with rudimentary small "penis" (Ж).

specimens) did not allow us to conduct a serious statistical analysis with a high probability of confirming or disproving this opinion about the "intermediate" place of imposex females of *Rapana* between typical female and male sexual groups in NWBS ecological conditions.

3.3 The first record of the biphallia phenomenon in male rapa whelk in the Black Sea

One of the varieties of imposexuality in anomalous females of Gastropoda is the appearance of more than one penial structure known as "double penis" or "biphallia." The first record of it was made for imposex females of *Leucozonia nassa* (Caenogastropoda: Fasciolariidae) in southeast Brazil [29]. However, similar deviations in the development of outward copulative organs may occur in typical males

Figure 6.
General view of male rapa whelk Rapana venosa *with normal (left) and "double" penis (right) (A) and the normal male outward genital structure of the Rapana (left) and abnormal "double" penis (right) in mollusk found in stormy landings on the Odessa coast, northwestern Black Sea (Ukraine), January 11, 2008 (B).*

of the gastropods. So, an adult male specimen of *Rapana venosa* with a double penis was accidentally found in storm deposits on the Odessa coast (46°27′48″N; 30° 45′47″E) on January 11, 2008. The anomalous penis of this mollusk was divided approximately in the middle into two parts, which were similar in shape with typical "hooks" on the end (**Figure 6A**). The length of the main (basic) trunk of penis was about 20 mm, perpendicularly from the main trunk grew an "appendix" that did not exceed 10 mm length (**Figure 6B**).

The age of the whelk was more than 2 years; the height of its shell (H) was 64.4 mm and the diameter (D) was 44.7 mm. The total mass of the animal (M_1) was 44.23 g; the mass of its raw body (M_2) was 22.68 g, and shell mass (M_3) was 17.22 g. Before this finding, about 130 specimens of typical males of this mollusk from the northwestern part of the Black Sea were studied, without observing similar deviations in the penial structure. Therefore, the frequency of appearance of such anomaly in this area is less than 0.8% [30].

The cases of appearance of "imposex" in various marine gastropod species have been used in several areas around the world as a tool to monitor the contamination by organotin compounds. So far, imposex-related biphallic structures have been reported for females of *Leucozonia nassa* from Brazil [29] and *Hexaplex trunculus* from the Tunisian coast [31]. As the described case of finding the anomalous male, *Rapana venosa* has not been studied with the connection between the appearance of biphallia unit and the marine water quality in the region; the author can only state the conclusion of same analogy.

4. Conclusions

The predatory marine gastropod *Rapana venosa* (Valenciennes, 1846) in the northwestern Black Sea (NWBS) has certain morphometric features in contrast to mollusks from the southeastern Crimean coast. Its shell is more elongated, thinner, and less massive. As a result, the relation of the total animal mass (M_1) and the soft body raw mass (M_2) and shell mass (M_3) in the NWBS area is more evident than in the Crimean populations: $M_1/M_2 = 2.5 \pm 0.03$ and 3.6 ± 0.04 and $M_1/M_3 = 1.9 \pm 0.01$ and 1.5 ± 0.01, respectively. The weight-size differences of the *Rapana* from NWBS may confirm certain morphometrical changes of the animals in different conditions of habitats in this area: low salinity and mineralization of the marine environment due to the strong freshwater influence of the river runoff (Danube, Dniester, South

Bug, and Dnieper), small depths in the coastal areas, etc. The evident morphometric features of *Rapana* in NWBS, the imposex appearance in females and "biphallia phenomenon" in males' mollusks from sea waters subject to anthropogenic pollution, make it possible to consider them as a promising object for further study as an indicator of the ecological state of marine coastal areas.

Acknowledgements

The author wishes to particularly thank the researchers of the Institute of Marine Biology (formerly the Odessa Branch of the Institute of Biology of the Southern Seas) National Academy of Sciences of Ukraine: A.P. Kurakin, for diving and collecting the live mollusks; V.V. Adobovsky, for his help in collecting the empty shells from the after-storm ejects on the Odessa coast; and Dr. A.Yu. Varigin, who found and kindly presented the anomalous mollusk with "double penis."

Author details

Igor A. Govorin
Institute of Marine Biology, National Academy of Sciences of Ukraine, Odessa, Ukraine

*Address all correspondence to: ia-govorin@ukr.net

References

[1] Golikov AN, Starobogatov YaI. Gastropod mollusks. Key for Black and Azov Seas fauna. In: Klass Briykhonogii Molluski—Gastropoda. Opredelitel' Fauny Chernogo I Azovskogo Morej. 3rd ed. Kiev: Naukova Dumka; 1972. pp. 65-166 [in Russian]

[2] Drapkin EI. A new mollusk in the Black Sea [Novyj mollusk v Chernom morie]. Priroda. 1953;**9**:92-95 [in Russian]

[3] Goncharov AD. *Rapana* from the northwestern coast of the Black Sea [Rapana u severo-zapadnogo poberezja Chernogo moria]. Hydrobiological Journal. 1977;**13**(3):20-31 [in Russian]

[4] Chukhchin VD. Ecology of the Gastropod Mollusks of the Black Sea [Ecologia Brykhonogikh Molluskov Chernogo Moria]. Academy of Sciences of USSR. Kiev: Naukova Dumka; 1984. 176 p [in Russian]

[5] Alyakrinskaya IO. On some morphophysiogical parameters of mollusks [Nekotoryje morfofiziologiczeskie pokazateli molluskov]. Hydrobiological Journal. 1996;**32**(1):72-83 [in Russian]

[6] Ksyunin GA. Age-size and Sexual Structure of *Rapana thomasiana* from Two Populations from Black Sea stocks [Razmerno-vozrastnaja I polovaja struktura dvuch poselenij chernomorskoj populatsii *Rapana thomasiana*]. Rybnoye Gospodarstvo Ukrainy. 2000;**5**:4-6 [in Russian]

[7] Frolenko LN, Studenikina EI, Golovkina EM. On the state of the populations of *Rapana thomasiana* in the northeastern part of the Black Sea [O sostojanii populatcii rapany *Rapana thomasiana* v severo-vostochnoj chasti Chornogo moria]. In: Mater. 2th International Conference "Marine Coastal Ecosystems. Algae, Invertebrates and Products of its Utilization". Moscow: VNIRO Publ. Agency; 2005. pp. 101-104 [in Russian]

[8] Kos'yan AR. Comparative analysis of *Rapana venosa* (Valenciennes, 1846) from different biotopes of the Black Sea based on its morphological characteristics. Oceanology. 2013;**53**(1):47-53. DOI: 10.1134/S0001437013010074

[9] Bilecik N. Distribution of Sea Snail Rapana Venosa in the Turkish Coast of Black Sea and its Impact on Black Sea Fisheries. Vol. 1, Seri B. Bodrum, Turkey: Directorate of Fisheries Institute Publ; 1990. 32 p

[10] Celik O, Samsun O. Investigation of the catch amount and the catch composition of dredges with various design features. Journal of Fisheries and Aquatic Science. 1996;**13**(3–4):259-272

[11] Gomoiu M-T. Some ecologic data on the gastropod *Rapana thomasiana* Crosse along the Romanian Black Sea shore. Cercetari Marine, IRCM. 1972;**4**: 169-180

[12] Micu S, Kelemen B, Mustata G. Current distribution and shell morphotypes of *Rapana venosa* (Valenciennes, 1846) in the Agigea 4 m littoral. Analele Ştiinţifice ale Universităţii "Al. I. Cuza" Iaşi, Serie Biologie Animală. 2008;**54**:185-189

[13] Smagovich K. Polymorphism and anomalous shells in juveniles of Rapana thomasiana Crosse, 1861 (Gastropoda: Prosobranchia: Neogastropoda) from the Black Sea Zeszyty Naukove Akademii Górniczo-Hutniczej im. S. Staszica. Folia Malacologica. 1989; **3**(1216):149-158

[14] Konsulova TH. Mussel *Mytilus galloprovincialis* Lam. (Bivalvia) natural resources along the northern Bulgarian

Black Sea coast in relation to *Rapana thomasiana* Crossa (Gastropoda) distribution. In: Trudy Instituta Oceanologii. Vol. 1. Varna: Bulgarian Academy of Sciences Publ; 1992. pp. 104-109

[15] Govorin IA, Kurakin AP. Finding of *Rapana thomasiana* in the near Danube Delta area of the Black Sea [Nakhodka rapany *Rapana thomasiana* v pridunajskom rajonie severo-zapadnoj chasti Chernogo moria]. Ecologiya Morya. 2005;**69**:18-19 [in Russian]

[16] Krylova TF, Globina ON. The use of Rapana from the Black Sea as food-stuffs [Ispol'zovanie chernomorskoj rapany na pischevyje tceli]. In: Ratsionalnoye ispol'zovanie syrievych resursov severo-zapadnoi chasti Chernogo morya: Thezy Dokladov Nauchno-prakt. Konference; 28–29 December 1984; Odessa; 1984. pp. 46-48 [in Russian]

[17] Project "Seas of the USSR": Hydrometeorology and hydrochemistry of the USSR seas. Vol. 4: Black Sea. Issue 1: Hydrometeorological conditions [Projekt "Morja SSSR". Gidrometeorologia i gidrokhimia morej SSSR. T. IV. Chernoje morie. Vyp. 1. Gidrometeorologicheskie uslovia]. St.-Peterburg: Gidrometeoizdat; 1991. 430 p [in Russian]

[18] Bondarev IP. Ecomorphological analyses of marine mollusks' shell thickness of Rapana venosa (Valenciennes, 1846) (Gastropoda: Muricidae). International Journal of Marine Science. 2013;**3**(45):368-388. DOI: 10.5376/ijms.2013.03.0045

[19] Kurakin AP, Govorin IA. Rate of *Mytilus galloprovincialis* consumption by *Rapana venosa* in the Northwestern part of the Black Sea. Hydrobiological Journal. 2011;**47**(6):15-22. DOI: 10.1615/HydrobJ.v47.i.6.20

[20] Govorin IA. Use of the morphometric characteristics of empty shells of Rapana thomasiana (Mollusca, Gastropoda), found in after storm ejects, for the determination of lifetime mass indices of mollusks [Ispol'zovanie morfometricheskikh kharakteristik rakovin *Rapana thomasiana* (Mollusca, Gastropoda) iz beregovykh shtormovykh vybrosov dla opredelenija prizhyznennykh pokazatelej massy molluskov]. Vestnik Zoologii. 2013;**47**(3):218 [in Russian]

[21] Foale S. An evaluation of the potential of gastropod imposex as a bioindicator of tributyltin pollution in Port Phillip Bay, Victoria. Marine Pollution Bulletin. 1993;**26**(10):546-552. DOI: 10.1016/0026-326X(93)90404-8

[22] Oehlmann J, Bauer B, Minchin D, Schulte-Oehlmann U, Fioroni P, Markert B. Imposex in *Nucella lapillus* and intersex in *Littorina littorea*: Interspecific comparison of two TBT-induced effects and their geographical uniformity. Hydrobiologia (Heidelberg). 1998;**378**:199-213

[23] Gibson CP, Wilson SP. Imposex still evident in eastern Australia 10 years after tributyltin restrictions. Marine Environmental Research. 2003;**55**(2): 101-112. DOI: 10.1016/S0141-1136(02)00097-1

[24] Neves Raquel AF, Valentin JL, Figueiredo GM. First record of imposex in Heleobia australis (Caenogastropoda: Cochliopidae). Journal of Molluscan Studies. 2013;**79**(1):82-85. DOI: 10.1093/mollus/eys034

[25] Savini D, Castellazzi M, Occhipinti-Ambrogi A. Observations on imposex on the alien invasive gastropod *Rapana venosa* (Valenciennes, 1846) in the northern Adriatic Sea. Biologia Marina Mediterranea. 2005;**12**(1):714-716

[26] Mann R, Harding JM, Westcott E. Occurrence of imposex and seasonal

patterns of gametogenesis in the invading veined rapa whelk *Rapana venosa* from Chesapeake Bay, USA. Marine Ecology Progress Series. 2006;**310**:129-138

[27] Castro IB, Fillman G. High tributylin and imposex levels in the commercial muricid *Thais chocolata* from two Peruvian harbor areas. Environmental Toxicology and Chemistry. 2012;**31**: 955-960

[28] Govorin IA. A phenomenon of female masculinization in *Rapana thomasiana* (Mollusca, Gastropoda) in the Northwestern part of the Black Sea. Vestnik Zoologii. 2007;**41**(3):212 [in Russian]

[29] Meirelles CAO, Castro IB, Pinheiro JCL. A first record of biphallia in imposexed female of Leucozonia nassa (Caenogastropoda: Fasciolariidae). JMBA2, Biodiversity Records. 2007:1-4

[30] Govorin IA. First record of the biphallia phenomenon in male rapa whelk *Rapana venosa* (Mollusca: Gastropoda) in the Black Sea. Mollusca. 2009;**27**(1):49-51

[31] Lahbib Y, Abidli S, El Menif NT. Imposex level and penis malformation in Hexaplex trunculus from the Tunisian coast. American Malacological Bulletin. 2008;**24**:79-89. DOI: 10.4003/0740-2783-24.1.79

Section 4

Immune System

Chapter 4

Immune Response of Molluscs

Hanan Al-Khalaifah and Afaf Al-Nasser

Abstract

In common with other invertebrates, molluscs are known to have internal immune response against foreign particles and organisms. The innate immunity of molluscs reflects the inherent non-specific response that provides the first line of defense. Anatomic barriers, phagocytic cells, and physiological components are the main elements of the innate immune response in molluscs. It is composed of both cellular and humoral elements. The cellular components are the circulating hemocytes. Small invaders are eliminated by the phagocytic hemocytes, while large invaders are eliminated by encapsulation. The ingested foreign particles are then hemolyzed by the action of certain toxic enzymes that catalyze oxidative burst reactions capable of killing pathogens and foreign invaders. Humoral components of molluscan immunity involve nitric oxide, lysozyme activity, lectins, and the phenyloxidase system. The current chapter sheds light on the elements of the molluscan innate immune system and presents a case study of the immune response of *Lymnaea stagnalis* mollusc against *Chaetogaster limnaei* parasite. The effect of the parasite on some humoral immune response parameters such as nitric oxide, phenol oxidase, and lysozyme production was investigated. In conclusion, the snail *Lymnaea stagnalis* exerts humoral immune response against *Chaetogaster limnaei* parasite. However, this response is insufficient to eliminate the parasite.

Keywords: *Chaetogaster limnaei*, humoral immune response, *Lymnaea stagnalis*, molluscs

1. Introduction

1.1 *Lymnaea stagnalis* in the animal kingdom

Molluscs are invertebrates that form one of the largest groups in the animal kingdom, with more than 100,000 known species. A wide range of molluscs including families of Pulmonata, Gastropoda, Planorbidae, and Lymnaeidae are intermediate hosts of trematodes.

Gastropods or snails are asymmetrical molluscs with a well-developed foot and radula. The visceral mass is spirally coiled. They occur in seas, freshwater, and terrestrial environments. The class Gastropoda is divided into three subclasses: Prosobranchia, Opisthobranchia, and Pulmonata. The snail *Lymnaea stagnalis* belongs to the family Lymnaeidae under the order Basommatophora, class Pulmonata. This snail is a freshwater scavenger that slides along the bottoms of ponds, lakes, and marshes looking for food. It is relatively large species with slender and sharply pointed spires. They can be found in North America, Europe, Asia, Algeria, and Morocco [1]. The head of the snail is on the front part of the foot that is used to glide on mucus slime produced from its own body. Pulling the foot inside the shell is one way of protection against predators. The shell walls are delicate and fairly transparent. The body is yellowish-gray in color, with a special adaptation of the mantle that allows it to take in air. Eggs are laid from April till the beginning of October and hatch after 10 days. Snails can reproduce when they are 3 months and can live for 6 years. They feed on algae, plants, other snails, and insects. The common name of this snail is the Great Pond Snail in the United Kingdom. This is one of the biggest freshwater snails with a height of 45–60 mm and a width of 20–30 mm [1]. **Plate 1** shows a *Lymnaea stagnalis* snail on vegetation [2].

1.2 Major parasitic infections associated with *Lymnaea stagnalis*

Lymnaea stagnalis is known to be an intermediate host of several parasites. Examples in Europe are *Fasciola hepatica*, *Trichobilharzia ocellata*, *Haplometra cylindracea*, and *Schistosomatium douthitti*.

Trichobilharzia ocellata is a trematode parasite of birds that causes what is known as the cercarial dermatitis of humans (swimmer's itch). This phenomenon has a high prevalence in Europe. The cercariae cannot develop into adults in human, but they cause erythema and urticaria of the invaded area, and a hypersensitivity reaction characterized by intense itching that may last several days up to 3 weeks. It can be caused by dead or dying larvae in the skin. **Plate 2(a)** shows a cercaria of *T. ocellata* and **Plate 2(b)** shows the swimmer's itch caused by penetration of the cercariae [3].

Haplometra cylindracea is a trematode that infects the frogs' lungs. It feeds on the blood that is supplied to the lungs. The xiphidiocercariae encyst in insects. Molluscan intermediate hosts could be: *Lymnaea stagnalis*, *L. ovata*, or *L. palustris*. This parasite is known to be non-pathogenic to man.

Schistosomatium douthitti is known as the rodent schistosome. It is also non-pathogenic to man. The percentage of natural infection with this parasite is only 3–7%. The natural definitive hosts of this parasite in North America are rodents. **Plate 3(a)** shows a cercaria of *H. cylindracea* and **Plate 3(b)** shows a cercaria of *S. douthitti* [4].

In addition, some annelids are found associated with the mantle area of *Lymnaea stagnali*. For example, *Chaetogaster limnaei* is a parasite of freshwater snails including *Lymnaea stagnali*. It belongs to the annelid family, Naididae, which is considered to be an ecologically diverse family of worms common in both running and standing waters. Reproduction occurs predominantly by paratomy, asexual reproduction

Plate 1.
A Lymnaea stagnalis *snail on vegetation.*

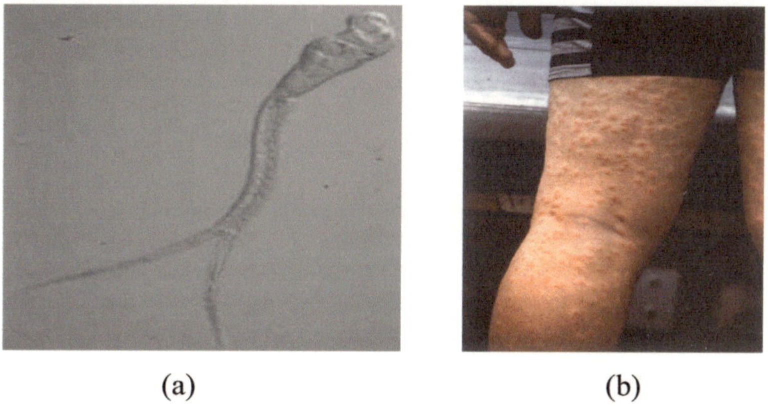

Plate 2.
(a) A cercaria of T. ocellata, *and (b) the swimmer's itch caused by penetration of the cercariae.*

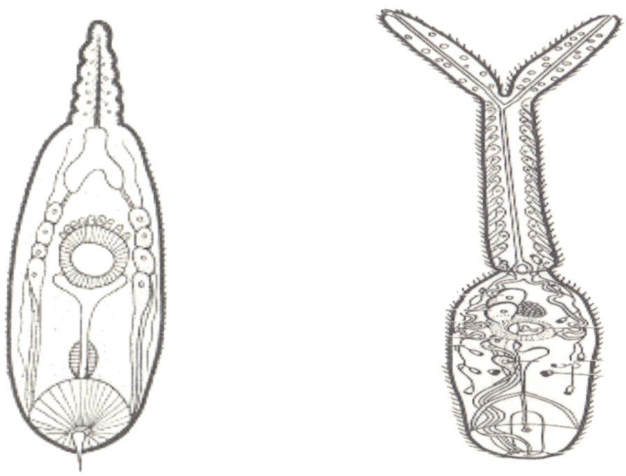

Plate 3.
(a) A cercaria of H. cylindracea, *and (b) a cercaria of* S. douthitti.

in which new organs are developed before the animal divides into two or more parts. In most cases, this organism appears to be commensalistic and does not cause high degree of mortality to its host, although it may affect the reproducing capacity by affecting the gonads. The length of the parasite ranges from 7 to 20 mm [5]. **Plate 4** shows *Chaetogaster limnaei* [6].

In common with other invertebrates, molluscs are known to have internal immune response against foreign particles and organisms. The innate immunity of molluscs reflects the inherent non-specific response that provides the first line of defense. Anatomic barriers, phagocytic cells, and physiological components are the main elements of the innate immune response. It is composed of both cellular and humoral elements. The cellular components are the circulating phagocytic hemocytes. Small invaders are eliminated by phagocytic hemocytes, while large invaders are eliminated by encapsulation. The ingested foreign particles are then hemolyzed by the action of certain toxic enzymes that catalyze oxidative burst reactions capable of killing pathogens and foreign invaders. Humoral components of molluscan immunity involve nitric oxide, lysozyme activity, lectins, and the phenyloxidase system [7, 8].

Plate 4.
Chaetogaster limnaei.

2. Innate response of molluscs

2.1 Anatomic barriers

Shells of molluscs act as physical barriers that prevent some pathogens from penetrating into the host's body. Mollusc shells are composed of organominerals with different structures and compositions depending on the species. Most of these shells are composed of an aragonite nacreous and a calcite prismatic layer [9]. The layers of the skin are tightly packed and the acidic environment in these layers is not suitable for the growth of most micro-organisms. The mucous membrane that lines the tissues and tracts contains normal micro-flora that competes with pathogenic micro-organisms for essential nutrients and attachment sites on these layers. The mucosal barrier may also act as a trap for antigens that are then inactivated by phage encapsulation by hemocytes. In addition, the movement of the outer cells expels the trapped pathogens to the outside environment. The flushing action of secretions helps to prevent infection of certain pathogens [10].

2.2 Key cells in innate immunity

There are immune cells in molluscs equivalent to the white blood cells in higher animals that play a role in the non-specific immune response. These cells are called hemocytes. Most of these cells are capable of engulfing extracellular particles by phagocytosis, endocytosis, and encapsulation. Others can produce substances that play a role in the killing process of pathogens. Circulating hemocytes proliferate and differentiate after exposure to pathogens or foreign particles. It was noticed that circulating hemoblasts are present at various phases of mitosis [11, 12].

The role of these phagocytic cells in the hemolymph of gastropods is considered to be the first line of defense system against invading or established organisms and particles. These hemocytes are circulating within the hemolymph as well as residing in the connective tissues. There are also cells lining the hemolymph spaces and are able to trap micro-organisms. Other cells are the fixed phagocytic reticulum cells that are fixed in the tissues by fibrils [13, 14].

A lot of studies have been done to determine the types and numbers of hemocytes in molluscs. This is considered to be a debating question because these hemocytes are morphologically and functionally heterogeneous. In addition to phagocytosis, these cells participate in digestion, excretion, healing of wounds, shell repair, transport, and encapsulation [14]. Classification of mollusc's hemocytes depends mainly on morphological criteria like cell size, nucleus to cytoplasm ratio, shape of nucleus, and the presence of granules in the cytoplasm using light microscopy. Studies have revealed the presence of two distinct types of hemocytes. These are the granulocytes and the agranulocytes. Acidophilic and basophilic granulocytes and slightly granular cells were also observed [15, 16]. The agranulocytes are relatively small cells and are also called hyalinocytes, they are less than 8 μm in size; while the granulocytes are larger spreading cells with pseudopodia and a polymorphic nucleus. The majority of cells in the hemolymph of molluscs are large spreading cells. Interestingly, there are intermediate stages between the round and the spreading cells in terms of morphology. Some scientists have suggested that the small and round cells are young cells and the larger spreading cells are old cells that are more capable of phagocytosis [17]. On the other hand, the small round cells in *L. stagnalis* do phagocytose, but not as much as the spreading cells. **Plate 5** shows hemocytes of *Lymnaea stagnalis* (A) free in the circulating hemolymph, while (B) shows hemocytes fixed in the tissue. **Plate 6** shows phase-contrast micrograph of *L. stagnalis* hemocytes, R is a round cell and the others are spreading hemocytes [17].

Tripp and Kent [18] have reported that 90% in vitro invading bacteria can be removed by hemocytes from molluscs within 24 h and 99% can be removed within 72 h. The glycolytic pathway in the cells was suggested to supply the hemocytes with the energy that is needed for phagocytosis, this was first proved by the work of Cheng [19] that has shown that hemocytes of the hard clam, *Mercenaria mercenaria*, use glucose/glycogen and produce lactate without an increase in the oxygen consumption.

2.3 Physiological elements of innate immunity in molluscs

In addition to the phagocytic hemocytes that are involved in the innate immunity, there are soluble physiological elements that play a role in defending the host from invading foreign substances and pathogens. Cellular and soluble elements coordinate in a sophisticated network of interactions to mount effective innate immune responses capable of eliminating foreign antigens. These soluble physiological elements include: nitric oxide, lysozyme activity, lectins, and the phenyloxidase system [7, 8].

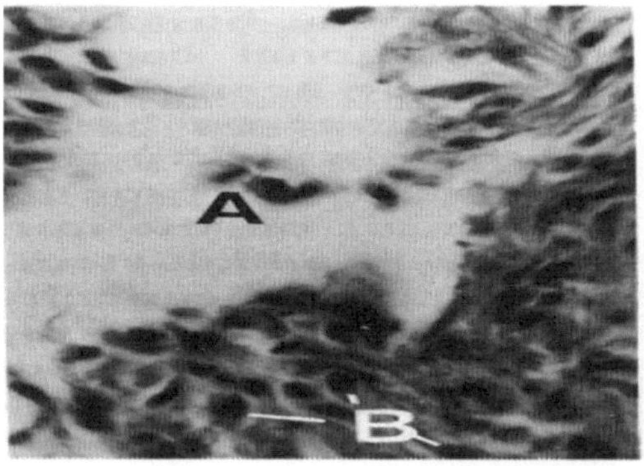

Plate 5.
Hemocytes of Lymnaea stagnalis, *(A) free in the circulating hemolymph and (B) hemocytes fixed in the tissue.*

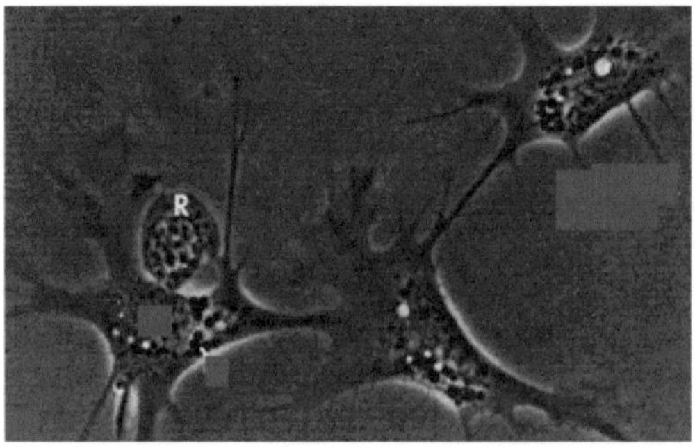

Plate 6.
Phase-contrast micrograph of L. stagnalis hemocyte. R is a round cell and the others are spreading hemocytes.

2.3.1 Complement system

The complement system is considered to be an important element of the innate immune system that also triggers the adaptive immunity in higher animals in the animal kingdom. Similar hemolytic complement-like activity was also reported against parasites and other pathogens in molluscs [20, 21].

Complement components are proteins and glycoproteins that are mainly synthesized in the liver. Other cells such as the tissue macrophages and the blood monocytes are involved in the complement production. These proteins are circulating in the blood in an inactive form. Once activated upon exposure to an antigen, they enter an enzymatic biochemical cascade that helps in the elimination of antigens by lysis of cells, opsonization, binding to specific complement receptors on cells of the immune system, and/or immune clearance of immune complexes [22–24].

There are three different pathways by which the complement cascade is activated, namely the classical pathway, the alternative pathway, and the mannan-binding lectin pathway. The classical pathway is antibody-dependent, so it is more related to the adaptive immunity. It is initiated by antigen-antibody binding to form immune complexes or when an antibody binds to an antigen on the surface of a pathogen. Activation of complement component 1, often simply called C1 (a protein of the immune system) then occurs when it binds to such antibodies of type IgM and some classes of type IgG. On the other hand, the alternative and the mannan-binding lectin pathways are more related to the innate immunity because they are antibody-independent. The former is initiated by complement component 3 (C3 complement), activation upon exposure of foreign substances on the cell wall of pathogens like bacteria, while mannan-binding lectin pathway is activated when a lectin binds to a mannose residue on the cell wall of pathogens such as *Salmonella*. This lectin is a serum acute phase protein that is produced as a result of the inflammatory response in the site of inflammation. The membrane attack complex (MAC) is produced upon activation of the complement system in the three pathways. This complex mediates lysis of the cell wall of bacterial pathogens. There are serum proteins as well as proteins on the surface of self-cells that control the activity of the complement system to ensure that host cells are not attacked [25–27].

2.3.2 Nitric oxide production

Many evidences have proven the presence of immunological factors in molluscs capable of working against infectious elements. Nitric oxide (NO) is a gaseous intercellular signaling molecule that has several functions in the cardiovascular, reproductive, nervous, and immune systems. Post infection by a pathogen, high levels of NO is synthesized in the immunocytes of the host in order to eliminate the pathogen or induce inflammation. Recently, NO has been reported to be involved in the immune response of several mollusc species such as scallop, clam, mussel, oyster, and snail. In addition to killing the invasive pathogens, high concentrations of NO can prove fatal to normal host cells as a result of strong cytotoxicity produced during the immune response. Modifying the NO concentration is crucial in order to maintain immune homeostasis and other physiological processes. Also, neuroregulation of NO is of utmost importance so as to prevent prolonged immune response [28].

Nitric oxide synthase was reported to play an important role in the phagocytic activity of hemocytes of molluscs such as *Mytilus galloprovincialis* and *Lymnaea stagnalis* [29–31]. *Lymnaea stagnalis* is a model organism for the study of the effects of environmental pollutants on immunological defense mechanisms, with one of the reasons being that its hemolymph collection is easy and does not require animal sacrifice. Most freshwater snails are dependent on cell-mediated cytotoxicity for the elimination of foreign materials. Phagocytosis plays a key role in the production of reactive oxygen species (ROS) known as the respiratory burst; this eliminates foreign particles and is the major defense mechanism in freshwater snails. Various stressors such as bacteria, viruses, parasites and xenobiotics affect the function of mollusc hemocytes. The elimination of ROS is important in order for the organism to survive. The antioxidant system which helps in this elimination includes superoxide dismutase (SOD), catalase (CAT), selenium-dependent glutathione peroxidase (Se-GPx), and glutathione reductase (GR). Glutathione (GSH) is an important scavenger of free radicals and helps in the maintenance of the redox status of proteinsulfhydryl groups. NO is produced by NO-synthases in hemocytes during phagocytosis and reacts with hydrogen peroxide to form peroxynitrite, a highly potent bactericide. NO may also act as an immunomodulator and mediates the effects of estrogen and opioids on immunity and inflammation [32].

2.3.3 Tyrosinase production

Tyrosinase is also considered to be a defense line system in the primary immune response of molluscs, it is involved in tanning and melanin formation. It is considered to be a non-self-recognition system because this enzymatic reaction can be triggered by minuscule amounts of molecules such as lipopolysaccharide, peptidoglycan and beta-1,3-glucans from micro-organisms. In *Lymnaea stagnalis* infected with the bird's schistosome *T. ocellata*, hemocytes were shown to increase their peroxidase activity between 2 and 8 weeks post infection [33, 34].

2.3.4 Lysosymes production

Lysozymes are enzymes that hydrolyze the 1,4-beta links between N-acetylmuramic acid and N-acetylglucosamine, they are known to destroy the cell walls of certain bacteria. They can be found in tears, saliva, egg whites, some plant tissues, and Sweat. Lysozymes have been recognized as a classic mollusc immune effector in innate immune system. It is a bacteriolytic enzyme produced from a range of organisms such as bacteria and bacteriophages to fungi, plants, and animals. They are characterized by their ability to bacterial peptidoglycan between two amino sugars, N-acetylmuramic acid and N-acetylglucosamine, which results in bacterial cell lysis and also has bactericidal and digestive abilities. In addition to this, they perform numerous other functions such as for growth stimulation, digestion, antiviral, anti-inflammatory, and are even associated with tumors. They are classified into six types based on their structural, catalytic and immunological characteristics. These are chicken-type (c-type), goose-type (g-type), plant, bacteria, T4 phage, and invertebrate type (i-type) lysozymes. The c-, g-, and i-type have been recorded in molluscs. They are the secreting type, which consist of signal peptide and 8, 6, and 14 cysteine residues, respectively. All three exhibit antibacterial and digestive ability and i-type has antifungi activity [35]. It was reported that the concentration of lysozymes in the hemolymph of molluscs changes as an immune response. Hemocytes of *L. stagnalis* were shown to have a selective bacteriostatic activity that is considered to be a defense factor against foreign materials. In addition of being destroying infectious elements, lysozymes also help in detection of the non-selfagents [33, 36].

2.3.5 Lectin production

Lectins binding phenomenon have been known since the turn of the nineteenth century, and have been widely used in the field of histochemistry for research and clinical purposes to detect specific carbohydrate structures and derivatives like glycolipids and glycoproteins in cells and tissues at the level of electron and light microscopy. They are also used for purification and isolation of carbohydrates and other specific cells like lymphocytes and bone marrow. In addition, lectins are very important tools in analyzing membrane structures of cells, glycosylation pathways, differentiation, cell division, growth and developmental changes, and mapping neuronal connections [37].

Lectin production is considered as an innate immune parameter against infection. These are non-enzymatic and non-immunoglobulin carbohydrate-binding proteins present in plants, bacteria, fungi, and animals which preferentially bind reversibly to specific carbohydrate structures, either free in solution or on cell surfaces. They are often classified based on saccharide-specificity. This specificity is usually defined by the monosaccharides or oligosaccharides that are best at inhibiting the agglutination or precipitation reactions caused by lectins. Lectins in the same category (e.g., galactose-specific lectins) show considerably different sugar-binding preferences. Interestingly, an increasing number of lectins which never show high affinity to simple saccharides have been found. Lectins can be detected by binding to fluorescent markers such as enzymes, biotin, and FITC and ^{125}Iodine [38–41].

Immune factors that consist of a fibrinogen domain (FBG) are now emerging as important components of the innate immune response of invertebrates. In recent times, some arthropod and mollusc model systems have contributed to the study of the functional role of fibrinogen-related proteins (FREPs) in invertebrate defense mechanism. FREPs have been identified in mussels, scallops, oysters, and gastropods. The FREPs of gastropod molluscs are interesting due to their unique structure, capacity for somatic diversification and their function in the immune system.

Gastropods and Digenean trematodes have shared a parasitic relationship for nearly 200 million years. These trematodes rely heavily on molluscs, mainly on gastropods to complete their larval development. They have developed mechanisms to evade and suppress specific snail immune processes so as to reproduce within the snail host. However, not every nematode can infect the snail species equally; this suggests that snails have developed mechanisms to fight trematode infections. A component of the snail immune response is BgFREPs; they are soluble lectin-like factors. Agglutinins have been discovered in gastropods such as *Helix pomatia* and *B. glbrata* and were involved in the recognition of non-self by binding to carbohydrate targets on hemocyte membranes of other species and pathogen-associated surfaces [42].

2.3.6 Phenoloxidase activity

Phenoloxidase (PO) is an enzyme that oxidizes phenols to produce melanin which plays an important role in egg production in gastropods, sclerotization of a new postmolt exoskeleton, and immunity of invertebrates. PO occurs as inactive precursors known as prophenoloxidses (proPO). They are activated by a proteolytic cascade system. Studies to investigate PO activity in *L. stagnalis* are limited. Their hemocytes play an important role in internal defense; they are involved in wound healing, encapsulation, and phagocytosis. They also contain several lysosomal enzymes such as peroxidase that are involved in intracellular killing. However, data in this aspect is lacking. Therefore, a study was taken to investigate the PO activity of *L. stagnalis*. The snails were collected from a lake in Russia and maintained under the desired laboratory conditions in 5-l aquaria containing dechlorinated tap water supplemented with mussel shell as a calcium source. Snails were fed lettuce daily and water was replaced once a week. Results showed that no PO activity was found. A low PO activity was seen in hemolymph without cells. On addition of a PO inhibitor, there was no effect on enzyme activity whereas hydrogen peroxide increased it [43].

3. Immune response of the snail *Lymnaea stagnalis* against parasites: study

The current chapter sheds light on some parameters that are associated with the immune response that is resulting from the interaction between the pond snail *Lymnaea stagnalis* mollusc and the parasite *Chaetogaster limnaei*. These parameters include production of nitric oxide, phenol oxidase, and lysozymes as an innate immune response that results from the interaction between the snail and the parasite.

3.1 Material and methods

3.1.1 Collection of snails and parasites

A total of 150 *Lymnaea stagnalis* snails were collected from Ringley, a freshwater pond in Manchester, England. The snails were usually supplied in a batch of 37–50 at a time and were kept in a fresh pond water tank at 4°C in the refrigerator. The water was changed every 5 days to prevent deoxygenation and death of snails. Pesticide-free lettuce was supplied as a source of food. The parasites were collected by scraping the body of the snail with a dissecting needle under a Thomas Scientific binocular dissection microscope and a Petri dish half filled with pond water to avoid drying of the snail. These parasites were then identified using the identification key of British Aquatic Oligochaeta [44].

3.1.2 Bleeding of the snails

Stroking the end of a dissecting pin over the foot of the snail would cause the snail's body to shrink back into the shell. This would expel any excess water outside the shell. Then, a dissecting needle was used to remove a portion of the shell just between the first and the second ring of the shell. Then, the gray membrane was punctured by a dissecting needle. Once punctured, the hemolymph is released and can be collected by using an Eppendorf micropipette set at 100 μl. This hemolymph was then collected in sterile Eppendorf tubes and kept on an ice tray in order to avoid aggregation of the cells.

3.1.3 Staining the hemocytes with Giemsa's stain

Thin smears of the hemolymph were made on microscopic slides and were let to dry for 20 min. 70% ethanol was used to fix the cells. The smears were then washed with distilled water and were stained with Giemsa's stain diluted in phosphate-buffered saline (PBS) containing sodium chloride, sodium phosphate, potassium chloride, and potassium phosphate at pH 7.3. The purpose of PBS is to maintain the constant pH and osmolarity of the hemocytes. The slides were washed, dried, and examined under light microscope at different powers.

3.1.4 Lysozyme standard curve

2 g of purified agar were added to 200 ml of phosphate-buffered saline (PBS) that has a pH value of 6.2. To this was added 0.15 g of *Micrococcus luteus* bacteria that was crushed prior to addition in order to avoid clumping within the solution. A stirring plate and a magnetic stirrer were used to stir the solution for approximately 15 min. Then, the solution was autoclaved at 121°C and 15 psi. The flask is then cooled in a water bath for 15 min and then poured in to a set of Petri dishes with a thickness of approximately 5 mm. Then, the plates were let to solidify in a biological cabinet. Twelve wells were created in the agar plate using a flame sterilized borer and a suction pipe to remove the agar pieces.

Lysozyme standard solutions were prepared by dissolving 25 mg of frozen-lysozyme in 5 mls of PBS at pH 6.2 to give a concentration of 5 mg/ml. Then, 12 serial two-fold dilutions were produced by removing 2.5 ml from the 5 mg/ml solution and add it to a tube containing 2.5 ml of PBS, this process was repeated until 12 serial dilutions were produced. These known concentrations of the lysozyme were dispensed into the wells that were produced in the agar plates, approximately 30 μl in a well. These plates were then incubated over night at 27°C. Areas of hydrolysis were observed around each well. The diameter of each zone was measured using a ruler and a lysozyme standard curve was plotted with the x-axis representing the lysozyme concentration and the y-axis representing the diameter (squared).

3.1.5 Lysozyme assay

4 agar plates were prepared as mentioned above, each containing 12 wells. 37 snails were bled and each hemolymph from each snail was placed into one of the wells. So, 37 wells in 4 agar plates were occupied. The plates were then incubated at 27°C overnight. The diameters of the hydrolysis zones were measured. The same assay was then repeated for 12 mucous samples and 12 parasite product samples.

3.1.6 Nitric oxide standard curve

A sodium nitrite standard solution was prepared. 100 μl of this solution was then placed in 2 wells at the top of a 96 wells enzymatic plate (ELISA plate) in duplicate. 50 μl of the culture medium was dispensed along the remaining 7 wells in duplicate. Then, serial two-fold dilutions of the standard solution were made going down the plate except for the last set of wells that were used as controls, they only contained culture medium without the standard solution. The culture medium is supplemented with 10% fetal calf serum.

Then, 50 μl of Griess reagent was added to all the wells. Once prepared, the tube should be inverted for 10 min and kept in the refrigerator. After addition of Griess reagent, the absorbance of the solutions in the wells should be read within 30 min using a plate reader with an absorbance filter of 570 nm. A nitric oxide standard curve was then plotted with the x-axis representing the nitrite concentration and the y-axis representing the absorbance.

3.1.7 Nitric oxide assay

Circular glass cover slips were placed in 15 wells of the 24-well tissue/cell culture plates using a pair of fine forceps. Ethanol was used to sterilize the cover slips. Poly-L-lysine solution was used to cover the cover slips in the wells, 100 μl in each well. This solution increases the hemocytes attachment to the cover slips. An incubation period of 1 h at 37°C was allowed in order for the poly-L-lysine to fix on the cover slips. After then, the excess poly-L-lysine solution was removed from the wells by Pasteur pipette and the wells were washed with sterile distilled water. Freshly collected hemolymph was then added to the wells, 50 μl in each well. The plate was then put in the refrigerator for 1 h to allow the hemocytes to attach themselves to the cover slips. The excess hemolymph was then removed by Pasteur pipette and the wells were washed by sterile distilled water.

The culture medium was then added to all the wells, 200 μl to each well. The plate was then incubated at 27°C for 1 h to enhance the growth of the hemocytes that can be observed under the light microscope. Approximately 10 live *C. limnaei* parasites were placed in to 3 wells (triplicate), another 10 dead *C. limnaei* were added in to another 3 wells. Parasite products were placed in to 3 wells. 3 wells containing the monolayers and the medium and another 3 wells containing only the medium were used as the controls. The plate was then incubated for 1 h at 27°C. Then, 200 μl of Griess reagent was added to all the wells. There should be a color change within a minute or so. Approximately 200 μl of the solution in each well was transferred into a 96-well enzymatic assay plate that was read using a plate reader set at 570 nm. This previous procedure was repeated using incubation periods of 3 h, 6 h, and overnight.

3.1.8 Phenol oxidase standard curve

Different concentrations of L-DOPA and fixed concentration of the tyrosinase enzyme were used to measure the initial rate of dopachrome formation at 475 nm wavelength. 5×10^{-3} mol dm^{-3} L-DOPA solution was used to prepare different solutions.

The 5×10^{-3} mol dm^{-3} DOPA solution was prepared by dissolving 0.0493 g of DOPA powder that has a molecular weight of 197.19 in 50 ml of the phosphate buffer saline. Then, 3 ml of each solution (tubes 1–8) were placed in plastic cuvette tubes and 0.1 ml of the tyrosinase solution was added. The enzyme solution was made by dissolving 10 mg of the enzyme (power) in 1 ml of the PBS. The tubes were quickly inverted once the enzyme was added and the absorbance was measured at 475 nm using a spectrophotometer. A graph was plotted with the x-axis representing the concentration of the L-DOPA solution and the y-axis representing the absorbance. From that graph, the most suitable concentration of the L-DOPA substrate to be used in the next step was determined. This concentration was shown to be the 0.1. Different concentrations of the enzyme were then used with the fixed concentration of the L-DOPA substrate that was determined in the previous step. 10 mg of the enzyme powder was dissolved in 1 ml of the phosphate buffer saline and serial dilutions of that stock solution were made. First, 0.2 ml of that stock solution was added into 1.8 ml of PBS to make a dilution of 1/10. Then, 0.2 ml of that solution was added again into 1.8 PBS to make a dilution of 1/100. Another 0.2 ml of the last solution was added into 1.8 of PBS to make 1/1000 dilution and so on until nine serial dilutions were made. Then, 1 ml of each enzyme dilution was added into 1 ml of the 0.1 concentration of the L-DOPA substrate solution. The tubes were then inverted quickly and the absorbance was read at 475 nm. A standard tyrosinase (phenol oxidase) curve was plotted and was used to determine the concentration of the enzyme in the samples.

3.1.9 Phenol oxidase assay

Monolayers of hemocytes were prepared as mentioned previously. Approximately 10 parasites were placed in to 3 wells (triplicate), another 10 dead parasites were added in to another 3 wells. Parasite products were placed in to 3 wells. 3 wells containing the monolayer and the medium and another 3 wells containing only the medium were used as the controls. The plate was then incubated for 1 h at 27°C. This previous procedure was repeated using incubation periods of 3 h, 6 h, and overnight. Then, 200 μl of the substrate solution (L-DOPA), 0.1 concentration of the 5×10^{-3} mol dm^{-3} solution was added to all the wells. The absorbance was then read at 475 nm wavelength using the ELIZA plate reader.

3.2 Results

All the 150 snails collected were found to be infected with the parasite *Chaetogaster limnaei*. The average number of hemocytes in 40 hemolymph samples was 2.489×10^6 cells.

3.2.1 Lysozyme assay

A standard curve of this experiment was first created (**Figure 1**). According to the equation of the standard curve, Y = 1.0158 X + 1.7312, the lysozyme concentration in the 37 hemolymph samples can be estimated. **Figure 2** shows the lysozyme concentration for the 37 hemolymph samples. Then, the lysozyme concentrations of the 12 mucous samples were estimated using the standard curve (**Figure 3**).

3.2.2 Nitric oxide assay

A standard curve was first plotted for the nitric oxide assay (**Figure 4**). Using the equation of the standard curve, y = 0.0014x + 0.0504, the x values that represent the nitrite concentration in the different samples were estimated. **Figure 5** shows the production of nitric oxide in response to live parasites, dead parasites, and parasite products.

3.2.3 Phenol oxidase assay

The absorbance of different concentrations of L-DOPA substrate is shown in **Figure 6** and the absorbance of different enzyme concentrations with fixed substrate concentration is shown in **Figure 7**. According to the equation of the standard

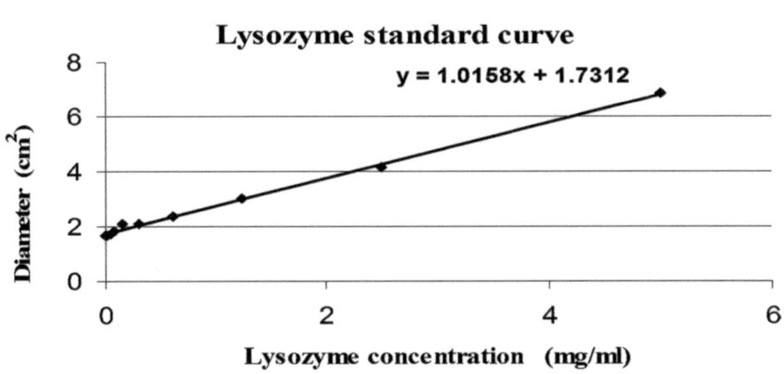

Figure 1.
Lysozyme standard curve.

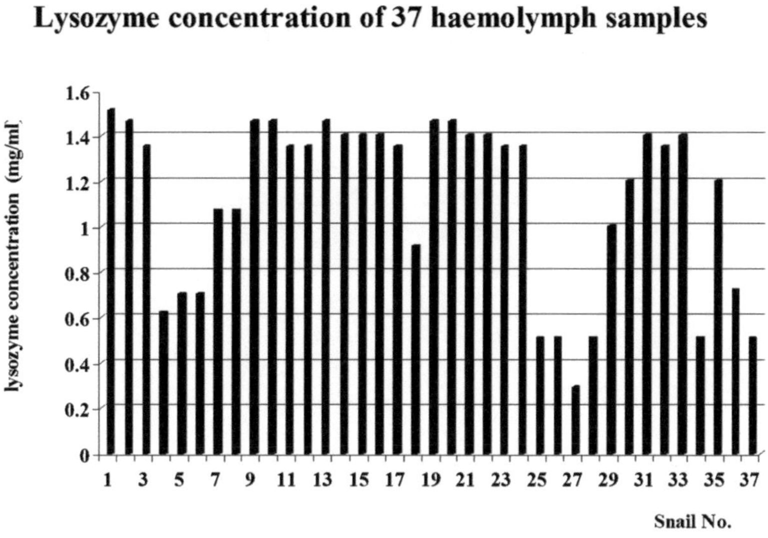

Figure 2.
Lysozyme concentrations of 37 hemolymph samples.

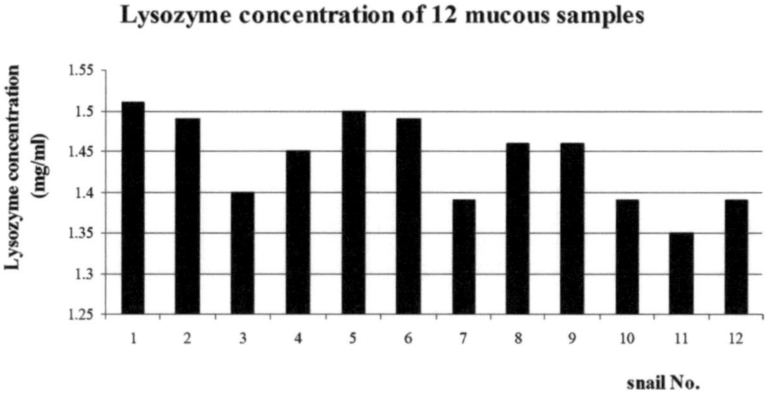

Figure 3.
Lysozyme concentrations of 12 mucous samples.

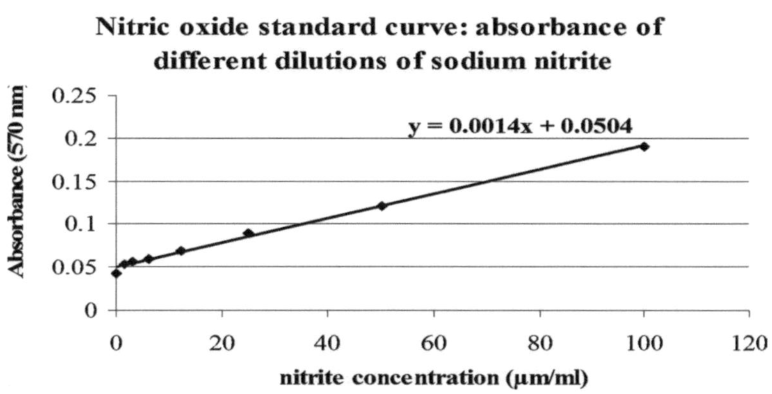

Figure 4.
Nitric oxide standard curve.

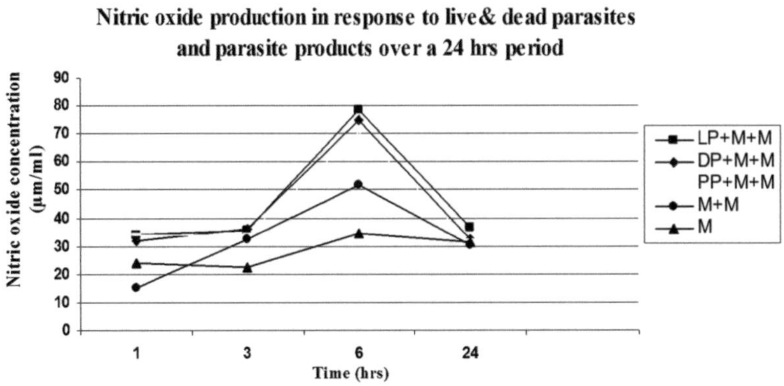

Figure 5.
Production of nitric oxide in response to live parasites, dead parasites, and parasite products. LP + M + M = live parasite, monolayer, and medium; DP + M + M = dead parasite, monolayer, and medium; PP + M + M = parasite product, monolayer, and medium; M + M = monolayer and medium; M = culture medium.

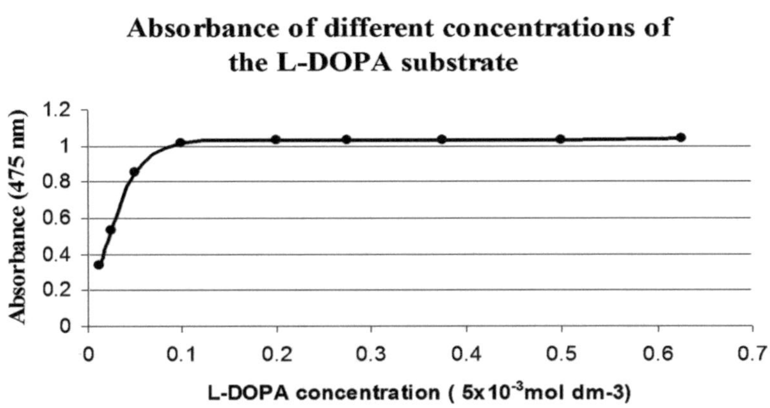

Figure 6.
Absorbance of different concentrations of the L-DOPA substrate.

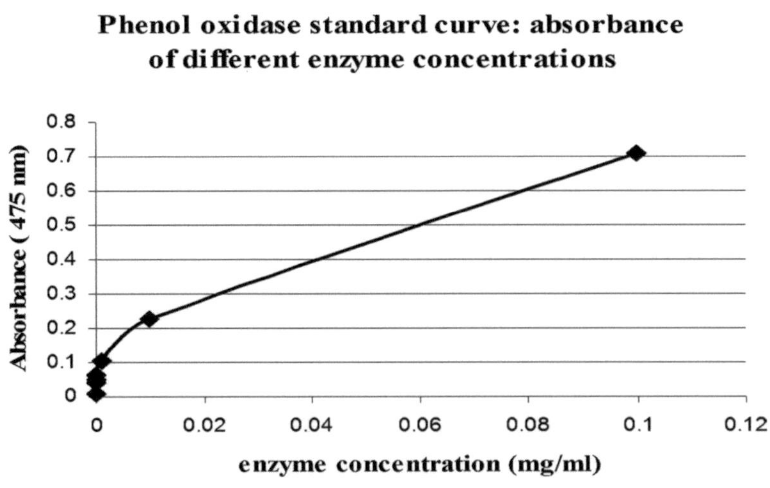

Figure 7.
Phenol oxidase standard curve: absorbance of different enzyme concentration.

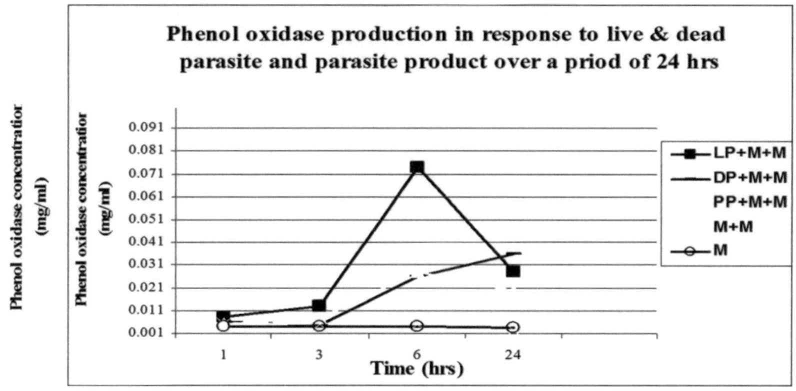

Figure 8.
Phenol oxidase production in response to live, dead parasites, and parasite products over a period of 24 h. LP + M + M = live parasite, monolayer, and medium; DP + M + M = dead parasite, monolayer, and medium; PP + M + M = parasite product, monolayer, and medium; M + M = monolayer and medium; M = culture medium.

curve, Y = 6.5223X + 0.0677, the x values that represent the phenol oxidase concentration in the different samples were estimated. **Figure 8** illustrates the previous results graphically.

4. Conclusion and discussion

The current work was conducted to shed light on some parameters that are associated with the immune response that is resulting from the interaction between the pond snail *Lymnaea stagnalis* and the parasite *Chaetogaster limnaei*. These parameters include production of nitric oxide, phenol oxidase, and lysozymes of the snail *L. stagnalis* as an innate immune response.

Lysozyme assay was carried out for hemolymph, mucous, and parasite products. Care was taken when mucous was collected in order not to contaminate mucous with hemolymph. The lysozyme standard curve was plotted on linear and logarithmic scale and the equation of that curve was used to estimate the lysozyme concentration in the unknown samples. In case of parasite product, zones of hydrolysis were not formed on the agar plates. This indicates that the parasite products did not contain any lysozymes. On the other hand, zones of hydrolysis were observed in the case of hemolymph and mucous samples. Accordingly, hemolymph and mucous of *Lymnaea stagnalis* did contain lysozymes that could have a role in hydrolysis of foreign particles invading the snail.

It can be noticed from **Figure 2** that there is a considerable fluctuation in the lysozyme concentration among the 37 hemolymph samples. This is due to variations in response to the presence of the bacteria in the agar. This may be due to the fact that some samples contain more hemolymph than hemocytes.

Figure 3 shows the lysozyme concentration in 12 mucous samples. As noticed, there is a little variation in the lysozyme concentrations among the 12 samples, indicating that different mucous samples contain different amounts of lysosomal enzymes. This is maybe due to differences in ages and sizes of the snails. Accordingly, it seems that the mucous represents the first defense line that prevents the establishment of the parasite in the snail's body.

Nitric oxide production results in response to live, dead parasites, and parasite products (**Figure 5**) do show considerable increase in nitric oxide production due to these inoculations. **Figures 5** and **6** are the same except for the control results (medium) that are shown in **Figure 5** and not in **Figure 6** in order for the reader to view the difference graphically. **Figure 5** shows that the culture medium does contain some nitric oxide, so, this value should be subtracted from the results of the samples to ensure that the resulting values are representing nitric oxide production from the hemocytes upon inoculation and not from the culture medium as shown in **Figure 5**. **Figures** 5 and **6** have shown that there is some nitric oxide produced in the case of cultured hemocytes monolayer and that the concentration of nitric oxide has increased upon inoculation of live, dead parasites, and parasite products. As shown in **Figure 5**, the optimum period of time for production of nitric oxide is 6 h postexposure to live parasites, followed by dead parasite, and then parasite products. In other words, live parasites induce more nitric oxide production than did dead parasites and parasite products. This indicates that viable parasites may produce antigens that are more efficient in initiating the production of nitric oxide. These antigens are less, in terms of efficiency or amount, in the case of dead parasites and parasite products. Another possibility is that some of the measured nitric oxide may be originated from the live parasite it's self.

It is noticed from **Figure 5** that the trend of the graphs in the case of different inoculations (live, dead parasites, and parasite products) is almost the same. After 1 and 3 h, there was little nitric oxide production, and then the concentration increased drastically after 6 h. The concentration of the nitric oxide decreased drastically after 24 h. This drop of nitric oxide concentration after 6 h may be due to the possibility that the hemocytes were viable and were able to produce high amounts of nitric oxide after 6 h. Then, the cells began to lose their viability and they were unable to produce the same amounts of that material. In addition, since different culture plates were used for different time intervals, the concentration of the hemocytes in the 6-h plates may be much more than the concentration in the overnight plate. Accordingly, more nitric oxide was measured in the 6-h plate than in the overnight plate. Moreover, this result may represent what is happening naturally in the snails' body. The concentration of nitric oxide at 6 h may be sufficient to get rid of the infection. After then, the concentration drops away because it already reaches the optimum and there is no need for more amounts. Another possibility is that some contaminants like detritus may be present in high amounts in the 6 h wells.

Results of the phenol oxidase assay are shown in **Figure 8**. The control results (medium and monolayer and medium by its self) were shown in the same figure and were not subtracted in order for the reader to view the difference graphically. It is noticed that the results of the controls, medium and monolayer, and medium alone are showing trace levels of phenol oxidase production (0.0040 mg/ml for the medium and 0.005 mg/ml for the medium and monolayer). It is also noticed that the production rate is steady throughout the experimental period. This indicates that there was no contamination with any other material that may induce sudden increase in the phenol oxidase production. It is shown in **Figure 8** that the optimum period of time for production of phenol oxidase is 6 h postexposure to the live parasites, followed by the dead parasites, and then the parasite products. In other words, live parasites induce more enzyme production than do dead parasites and parasite products. This indicates that viable parasites may produce antigens that are more efficient in initiating the production of phenol oxidase. These antigens are less in the case of dead parasites and parasite products or are not that efficient in stimulating the production of the enzyme. In the case of live parasites, there was little increase in the phenol oxidase production until 3 h. Then the production rate increased drastically to reach its maximum at 6 h postinoculation. After then, it decreased throughout the experimental period. This peak may be due to some contaminants that stimulate the hemocytes to produce more phenol oxidase in the 6 h wells. Also, the hemocytes maybe more viable and were able to produce high amounts of phenol oxidase at 6 h. Then, the cells began to loose their viability and they were unable to produce the same amounts of that enzyme at 24 h. In addition, since different culture plates were used for different time intervals, the concentration of the hemocytes in the 6-h plates may be much more than the concentration in the other plates. Accordingly, more phenol oxidase was measured in the 6-h plate than in the other plates. Another possibility is that this happens naturally in the snails' body. The concentration of phenol oxidase at 6 h may be sufficient and enough to get rid of the infection. After then, the concentration drops away because it already reaches the optimum and there is no need for more amounts.

It is shown in **Figure 8** that the phenol oxidase production in the case of the dead parasites increases until it reaches its maximum at 24 h. However, the amount produced is much more less than in the case of live parasites. This indicates the antigenic nature of the materials produced by the dead parasites is much less than those produced by the live organisms. The maximum amount produced did not drop away, this may be due to the possibility that this amount is still not sufficient to get rid of the dead parasites and more amounts are needed to be produced. If these samples are kept more than 24 h, a decrease in phenol oxidase production maybe observed.

The production of phenol oxidase in the case of parasite products has almost the same trend as in the case of dead parasites until 6 h postinoculation. After then, the enzyme production decreased. So, there is a slight peak for phenol oxidase production in the case of parasite products, as can be noticed in **Figure 8**. This peak maybe due to the same possibilities discussed earlier. So, phenol oxidase production results in response to live, dead parasites, and parasite products (**Figure 8**) do show a noticeable increase in enzyme production due to these inoculations.

In a similar study, in the first experiment, snails were randomly assigned into six different treatment groups. There was no difference in their sizes. The first group was not exposed to any immune elicitor, the second group was injected with 100 μl of snail saline, the third and fourth groups were injected with lyophilized *E. coli*. The fifth and sixth groups were injected with *Plagiochis* sp. Six hours post injections, the snails hemolymph was sampled for immunocompetence. Samples were also collected to measure PO and antibacterial activity. A second experiment was conducted by exposing the snails to opportunistic micro-organisms into two different water qualities. Statistical analyses were done using multivariate analysis of variance (MANOVA). Results showed that both, PO and antibacterial activity of snail hemolymph differed among the treatment groups. Wounding of the snail by injecting with snail saline increased PO activity of the hemolymph as compared to snails that did not receive any injection. Injecting with immune elicitors did not further increase the PO activity. Levels of PO activity decreased in those injected with *E. coli* wounding did not affect antibacterial activity. For the second experiment, the level of PO activity in snails maintained n micro-organism-rich water was higher compared to snails maintained in clean water [45].

5. Conclusion

In conclusion, lysozymes, nitric oxide, and phenol oxidase play an important role in the non-specific immune response against invading parasites and foreign particle in *Lymnaea stagnalis* snail. However, this immunity was not efficient to eliminate the parasite in the case study mentioned above.

Acknowledgements

The authors would like to extend their gratitude and appreciation to the management of the Kuwait Institute for Scientific Research for their continuous technical and financial support of scientific research.

Conflict of interest

There is no conflict of interest related to the current work.

Author details

Hanan Al-Khalaifah* and Afaf Al-Nasser
Environment and Life Sciences Research Centre, Kuwait Institute for Scientific Research, Safat, Kuwait

*Address all correspondence to: hkhalifa@kisr.edu.kw

References

[1] Abbaci K et al. Anatomical and histological characterization of the gametogenesis of *Radix balthica* (Linnaeus, 1758) in comparison with *Lymnaea stagnalis* (Linnaeus, 1758). Journal of Histology & Histopathology. 2017;**4**:5

[2] Lycos, http://members.lycos.co.uk/Mollusks/Schnecken/parasitismus/fasciola.html [Accessed on: 18 June, 2004]

[3] Wilfried H, Tina L, Kerstin G. http://www.biologie.unierlangen.de/parasit/contents/research/tricho.html [Accessed on: 18 June, 2004]

[4] Smyth JD. Introduction to Animal Parasitology. 2nd ed. UK: Hodder and Stoughton; 1981. pp. 184-203

[5] Jones RA et al. The prevalence and development of digenean parasites within their intermediate snail host, *Galba truncatula*, in a geographic area where the presence of *Calicophoron daubneyi* has recently been confirmed. Veterinary Parasitology. 2017;**240**:68-74

[6] Waegele H, et al. 2003. http://home.t-online.de/home/ghstanjek/mic1.htm [Accessed on: 20 June, 2004]

[7] Silva dos Santos F et al. Evaluation of the immune responses of the brown mussel *Perna perna* as indicators of fecal pollution. Fish & Shellfish Immunology. 2018;**80**:115-123

[8] Wang W et al. Pathogen-derived carbohydrate recognition in molluscs immune defense. International Journal of Molecular Sciences. 2018;**19**(3):721

[9] Dauphin Y et al. Structure and composition of *Unio pictorum* shell: Arguments for the diversity of the nacroprismatic arrangement in molluscs. Journal of Microscopy. 2018;**270**(2):156-169

[10] Skala V, Walker AJ, Horak P. Extracellular trap-like fiber release may not be a prominent defence response in snails: Evidence from three species of freshwater gastropod molluscs. Developmental & Comparative Immunology. 2018;**79**:137-141

[11] Cima F, Matozzo V. Proliferation and differentiation of circulating haemocytes of *Ruditapes philippinarum* as a response to bacterial challenge. Fish & Shellfish Immunology. 2018;**81**:73-82

[12] Mahapatra E et al. Sustaining immunity during starvation in bivalve mollusc: A costly affair. Tissue and Cell. 2017;**49**(2, Part B):239-248

[13] Grandiosa R et al. Effect of antiaggregants on the in vitro viability, cell count and stability of abalone (*Haliotis iris*) haemocytes. Fish & Shellfish Immunology. 2018;**78**:131-139

[14] Van der Knaap WPW, Meuleman EA, Sminia T. Alternations in the internal defence system of the pond snail *Lymnaea stagnalis* induced by infection with the schistosome *Trichobilharzia ocellata*. Parasitology Research. 1987;**73**:57-65

[15] Tame A et al. Morphological and functional characterization of hemocytes from two deep-sea vesicomyid clams *Phreagena okutanii* and *Abyssogena phaseoliformis*. Fish & Shellfish Immunology. 2018;**74**:281-294

[16] Cheng TC. Functional morphology and biochemistry of mollscan phogocytes. Annals of the New York Academy of Sciences. 1975;**266**:343-379

[17] Michel B. Immunity in Invertebrates. Berlin Heidelberg, Germany: Springer-Verlag; 1986. pp. 25-35

[18] Tripp MR, Kent VE. Studies on oyster cellular immunity. In Vitro. 1967;**3**:129-135

[19] Cheng TC. Aspects of substrate utilizing and energy requirement during molluscan phagocytosis. Journal of Invertebrate Pathology. 1976;**27**:263-268

[20] Niu D et al. Hemolytic reactions in the hemolymph of bivalve *Sinonovacula constricta* show complement-like activity. Fish & Shellfish Immunology. 2018;**79**:11-17

[21] Dubos M-P et al. Characterization of a tachykinin signalling system in the bivalve mollusc *Crassostrea gigas*. General and Comparative Endocrinology. 2018;**266**:110-118

[22] Bacon LD, Witter RL. Influence of B-haplotype on the relative efficacy of Marek's disease vaccines of different serotypes. Avian Diseases. 1993;**37**(1):53-59

[23] Powell PC. Fundamental concepts of avian immunology [French]. Point Veterinaire. 1983;**14**(70):57-66

[24] Goldsby RA et al. Immunology. 5th ed. Vol. xxiii. New York: W.H. Freeman; 2003. 576p

[25] Dorny P et al. Serum haemolytic complement levels in German Dahlem red chickens are affected by three major genes (naked neck, dwarf, frizzled) of tropical interest. Tropical Animal Health and Production. 2005;**37**(1):1-9

[26] Baelmans R et al. Different serum haemolytic complement levels in indigenous chickens from Benin, Bolivia, Cameroon, India and Tanzania. Tropical Animal Health and Production. 2004;**36**(8):731-742

[27] Frank SA. Immunology and Evolution of Infectious Disease. Vol. x. Princeton, N.J., Oxford: Princeton University Press; 2002. 348p

[28] Jiang Q et al. Mutual modulation between norepinephrine and nitric oxide in haemocytes during the mollusc immune response. Scientific Reports. 2014;**4**:6963

[29] Jiang G, Yu R, Zhou M. Studies on nitric oxide synthase activity in haemocytes of shrimps *Fenneropenaeus chinensis* and *Marsupenaeus japonicus* after white spot syndrome virus infection. Nitric Oxide: Biology and Chemistry/Official Journal of the Nitric Oxide Society. 2006;**14**(3):219-227

[30] Novas A et al. Nitric oxide release by hemocytes of the mussel *Mytilus galloprovincialis Lmk* was provoked by interleukin-2 but not by lipopolysaccharide. The International Journal of Biochemistry & Cell Biology. 2004;**36**(3):390-394

[31] Serfozo Z et al. Nitric oxide-coupled signaling in odor elicited molecular events in the olfactory center of the terrestrial snail, *Helix pomatia*. Cellular Signalling. 2017;**30**:67-81

[32] Gust M et al. Effects of short-term exposure to environmentally relevant concentrations of different pharmaceutical mixtures on the immune response of the pond snail *Lymnaea stagnalis*. Science of the Total Environment. 2013;**445**:210-218

[33] Carballal MJ et al. Enzymes involved in defense functions of hemocytes of mussel *Mytilus galloprovincialis*. Journal of Invertebrate Pathology. 1997;**70**(2):96-105

[34] Gopalakrishnan S et al. Modulation and interaction of immune-associated parameters with antioxidant in the immunocytes of crab *Scylla paramamosain* challenged with lipopolysaccharides. Evidence-based Complementary and Alternative Medicine. 2011;**2011**

[35] Jielian W et al. Characterization and roles of lysozyme in molluscs. ISJ—Invertebrate Survival Journal. 2017;**14**:432-442

[36] Matozzo V et al. Assessing the health status of farmed mussels (*Mytilus galloprovincialis*) through histological, microbiological and biomarker analyses. Journal of Invertebrate Pathology. 2018;**153**:165-179

[37] Bovi M et al. BEL β-trefoil: A novel lectin with antineoplastic properties in king bolete (*Boletus edulis*) mushrooms. Glycobiology. 2013;**23**(5):578-592

[38] Holmskov U, Thiel S, Jensenius JC. Collectins and ficolins: Humoral lectins of the innate immune defense. Annual Review of Immunology. 2003;**21**(1):547-578

[39] Rogers NC et al. Syk-dependent cytokine induction by dectin-1 reveals a novel pattern recognition pathway for C type lectins. Immunity. 2005;**22**(4):507-517

[40] Gabius H. Animal lectins. European Journal of Biochemistry. 1997;**243**(3):543-576

[41] Sharon N, Lis H. History of lectins: From hemagglutinins to biological recognition molecules. Glycobiology. 2004;**14**(11):53R-62R

[42] Gordy MA, Pila EA, Hanington PC. The role of fibrinogen-related proteins in the gastropod immune response. Fish & Shellfish Immunology. 2015;**46**(1):39-49

[43] Vorontsovaa Y et al. Do snails *Lymnaea stagnalis* have phenoloxidase activity in hemolymph. Invertebrate Survival Journal. 2015;**12**:5-12

[44] Brinkhurst RO. A guide for the identification of British aquatic Oligochaeta. Scientific publication. London: Freshwater Biological Association. Vol. 22. 1963. pp. 1-55

[45] Seppälä O, Leicht K. Activation of the immune defence of the freshwater snail *Lymnaea stagnalis* by different immune elicitors. Journal of Experimental Biology. 2013;**216**(15):2902-2907

Chapter 5

Survival Differences of *Vibrio vulnificus* and *Vibrio parahaemolyticus* Strains in Shellstock Oysters (*Crassostrea virginica*) from Harvest to Sale: A Risk Perspective

Violeta Pardío, Irma Wong, Leonardo Lizárraga, Karla López, Argel Flores, Guadalupe Barrera, Francisco Alarcón and Carlos Fernández

Abstract

As there is limited information on the risk for consuming market oysters contaminated with *V. vulnificus* and *V. parahaemolyticus*, the aim of this study was to estimate the risk associated with raw oyster consumption affected by contamination levels and temperature during postharvest and transportation. To evaluate the effect of the temperature during transportation from the Mandinga Lagoon to Mexico City on the growth of *V. vulnificus* and *V. parahaemolyticus*, a modified Gompertz model was fitted at ambient temperatures of 20.1, 25.6, and 24.4°C for 22 h in windy, dry, and rainy seasons, respectively. The risk was calculated using FDA/FAO/WHOv.2005 software. Results showed that the mean risk (cases per 100,000 servings) of a person acquiring *V. vulnificus vvha+/cvg*C infection by consuming raw oysters was 2.9×10^{-6}, 4.7×10^{-6}, and 4.3×10^{-6} during windy, dry, and rainy seasons, respectively. Risk for consuming oysters during windy season at-harvest contaminated with *V. parahaemolyticus tdh+* was 8×10^{-6} and 7.8×10^{-7} for consuming oysters at-market during rainy season contaminated with *V. parahaemolyticus tdh+* and *trh+*. These results suggest that maintaining temperatures above 20°C during oyster storage and transportation increases the risk of infections by pathogenic strains. The results provide a benchmark information to establish strategies to improve public health.

Keywords: *Crassostrea virginica*, *Vibrio*, harvest, market, season, survival, risk

1. Introduction

Vibrio vulnificus and *Vibrio parahaemolyticus* are the etiologic agents of seafood-associated fatalities worldwide. These Gram-negative, halophilic bacteria found naturally in marine and estuarine waters have the ability to cause lethal infections including primary septicemia, wound infection, and gastroenteritis associated with the consumption of raw or undercooked seafood, particularly oysters, throughout the world [1–4]. *V. vulnificus* is more frequently associated with wound infections, with a case fatality rate as high as 50% [5], particularly in individuals with predisposing conditions, including patients with chronic liver disease, immunodeficiency, iron storage disorders, end-stage renal disease, and diabetes mellitus [6]. Similarly, *V. parahaemolyticus* infection can cause diarrhea and septicemia that may be life-threatening to people having underlying medical conditions such as liver disease, diabetes, or immune disorders [7, 8]. The *tlh* (thermolabile hemolysin) gene is a species-specific marker for *V. parahaemolyticus*, while the *tdh* (thermostable direct hemolysin) and *trh* (thermostable-related hemolysin) genes are pathogenicity markers for *V. parahaemolyticus* [9]. The occurrence of *orf8* genes has been considered an additional virulence factor for *V. parahaemolyticus* [10, 11]. *V. vulnificus* includes three biotypes of which Biotype 1 is capable of producing fatal disease to humans due to consumption of raw seafood. Biotype 1 has been further divided into two genotypes, C and E. The gene *vcg* (virulence-correlated gene) has two alleles, *vcgC* and *vcgE*, representing clinical and environmental strains, respectively [11].

Vibrio vulnificus and *V. parahaemolyticus* are commonly reported in many countries around the world with high mortality rates [12]. In Mexico, *V. vulnificus* was isolated in 27% (39/143) of oyster samples collected from Pueblo Viejo Lagoon, located on the North Gulf Coast of Veracruz state, Mexico. Isolation rates were significantly higher in June ($P < 0.0002$) and *V. vulnificus* was found to prefer salinity conditions above 18‰ and temperatures above 24°C ($P < 0.001$) [13]. Meanwhile, *V. parahaemolyticus tdh+* incidence has been reported in raw oysters (44.0%) sold in Guadalajara, México, during the warm months ($P = 0.0038$) [14], and in oyster samples (8.7%) from Pueblo Viejo Lagoon in Tamaulipas, México as well; likewise, in the coastal zone of Tamaulipas, México, a 19.9% prevalence of *V. parahaemolyticus* in oysters was reported, which increased 18.3 times during summer months (July, August, and September) [15]. Our studies [16] revealed the highest mean densities of *V. parahaemolyticus tlh+, tdh+/trh, tdh/trh+* and *tdh+/trh+* during spring season at 2.57, 1.74, 0.36, and 0.40 $\log_{10}$ MPN/g, respectively, and *tdh+/orf8+* during winter season (0.90 $\log_{10}$ MPN/g) in oysters harvested from Mandinga Lagoon System (MLS) located on the coast of Veracruz, Mexico. *V. parahaemolyticus tlh +* densities were associated to salinity ($R^2 = 0.372$, $P < 0.022$), *tdh+/trh+* to turbidity ($R^2 = 0.597$, $P < 0.035$), and *orf8+* to temperature, salinity, and pH ($R^2 = 0.964$, $P < 0.001$) [16]. In this context, the exposure to salinity and temperature conditions regulate the dynamics of *V. vulnificus* and *V. parahaemolyticus* harboring potentially pathogenic genotypes within the oyster. This adaptive response of *V. vulnificus* and *V. parahaemolyticus* to seasonal environmental changes may lead to an increase in survival and virulence, threatening the seafood safety and increasing the risk of illness [16].

The American oyster (*Crassostrea virginica*) is one of the most popular bivalve mollusks, widely consumed in large quantities. In Veracruz state, oysters are harvested extensively within the oyster-producing areas found along the Mexican Gulf coast. The state of Veracruz is the primary oyster producer, harvesting 26,713 tons annually, which accounts for 43% of the national average annual production (61,996 t) [17]. They are sold alive in whole shell, shucked in fresh form or packaged, and refrigerated in polyethylene bags. According to the Mexican Norm [18], which provides guidelines for the sanitary control and commerce of shellfish in Mexico, shellstock oysters should be kept alive and adequately refrigerated to an internal body temperature of 7°C for 7 days at most to ensure safe consumption. Nevertheless, during transport and storage of raw oysters, adverse conditions (low oxygen levels, accumulation of waste, feeding interruption, and temperature abuse) favor recontamination and rapid deterioration [19]. *V. vulnificus* and *V. parahaemolyticus* can multiply in postharvest shellfish if they are held at temperatures >10°C [20, 21]. Although our previous studies have revealed a high prevalence of *V. parahaemolyticus* in oysters (*C. virginica*) in Veracruz, a relatively high proportion of oysters sold is not currently subjected to any postharvest process and is thus a health hazard. The MLS is an important area economically, where seafood production and consumption are common. It represents one of the most productive estuarine-lagoon systems in the Mexican Gulf of Mexico for year-round oyster harvesting with an oyster production of 306 t/y, resources that are supplying to seafood restaurants and oyster bars from nearby cities, mostly Veracruz—Boca del Río, and to Cancún and México City [22]. Because of the importance of raw oysters in gastronomy and economics, their microbial safety is of major interest. However, there is limited information on the loads of *V. vulnificus* and *V. parahaemolyticus* in oysters at market after long-distance transportation. Therefore, the aim of this study was to compare the seasonal survival ability of *V. vulnificus* and *V. parahaemolyticus* in shellstock oysters transported under ambient air and dry storage conditions from the MLS to a wholesale market in Mexico City, and to assess the risk as affected by storage and transportation conditions.

2. Materials and methods

2.1 Oyster collection and transportation

Six-specimen collections were performed from the same lot of oysters during dry, rainy, and windy seasons from January to December 2012 in two different sites: (1) at the oyster harvesting bank Mata Grande (Pescadores Unidos Union producer) in the MLS at 08:00 am by divers and (2) directly from the customer at the *Central de Abasto* in Mexico City at 08:00 am next morning, where oysters from this producer are sold. This is one of the most important wholesale seafood markets in Mexico City. Mata Grande oyster bank is located close to mangrove islands in Mandinga Grande lagoon (**Figure 1**). The MLS is located in southern state of Veracruz, Mexico, flows parallel to the northwestern coastline of the Gulf of Mexico, between 19°02′ N and 96°06′ W in Alvarado, Veracruz. MLS is formed by the confluence of the river Jamapa, and effluents of Huatusco, Cotaxtla, Totolapan rivers, ending in the Gulf of Mexico by the Boca del Rio, close to Veracruz City. It is

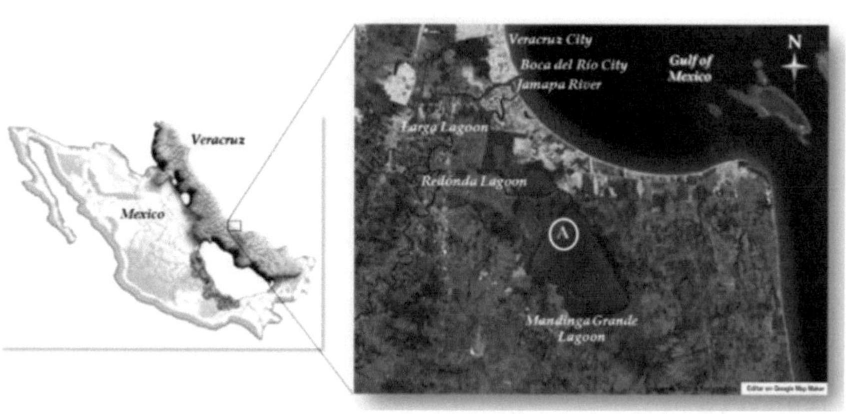

Figure 1.
Location of the study region and map of the MLS. Site of oyster samples collection monitored during dry, rainy, and windy seasons: bank A Mata Grande located close to mangrove islands 19° 01′ 53.8″N and 96° 04′ 23.1″W.

a shallow (1–3 m depth) tropical lagoon connected to the sea by a long and narrow deeper channel through the Jamapa River. This lagoon system consists of four lagoons (Conchal, Larga, Chica, and Grande) and flooded zones and cover an area of 3250 ha. The dry season occurs from March to June, the rainy season occurs from July to October, and the windy season from November to February when the MLS is affected by high-velocity northern winds (90–129 m/s) [16, 23].

Producers harvested the oysters at 08:00 am in the morning and stored at ambient temperature in a storage room until 18:00 pm when oysters were loaded in sacks for transport. Oyster sacks were transported stacked on a nonrefrigerated box truck overnight by road, arriving to the *Central de Abasto* market in Mexico City at 03:00 am. Product was delivered to customer at 6:00 am and samples were collected at 08:00 am. The average transit time was 22 h supply-chain from MLS to Mexico City. This producer transports oysters at ambient temperature and no records were available to document postharvest temperature exposure, which creates increased opportunity for temperature abuse. Therefore, specific practices and sampling points were selected based on those that are currently in use. A total of 80 legal-sized [24] live shellshock oysters were immediately transported to the laboratory according to Mexican Minister of Health approved method NOM-109-SSA1-1994 [25]. Dead animals were discarded, and the remaining oysters were scrubbed and rinsed under cold running tap water to remove debris and attached algae.

2.2 Bacteriological analysis

Within 2 h of collection, oysters were shucked, and meats and intravalvular liquids were pooled under aseptic conditions. Oyster samples were analyzed according to the protocol of Lizárraga-Partida et al. [26] modified. *V. vulnificus* and *V. parahaemolyticus* quantification was performed following the same most probable number-polymerase chain reaction (MPN-PCR) procedure described previously [16, 27], briefly: a 200 g of oyster sample (150 g of meat and 50 g of intravalvular fluid) were mixed with 200 mL phosphate-buffered saline (PBS) and blended for 120 s to make a 1:1 dilution. The shellfish homogenate was added to alkaline peptone water in a three-tube MPN dilution series prepared up to $1{:}10^4$ according to the standard three-tube MPN procedure. The tubes were incubated at 35°C for 24 h. After incubation, DNA was extracted from each positive APW tube showing growth and then purified. The densities of *V. vulnificus* and *V. parahaemolyticus* strains were calculated using positive results by PCR, employing the most probable number (MPN) tables. Simultaneously, one loopful from the top 1 cm of each positive broth tube from the MPN method categorized as positive for *V. vulnificus vvha* + and *V. parahaemolyticus tlh*+ based on the DNA amplification results was streaked onto CHROMagar™ Vibrio (CHROMagar Microbiology, Paris, France). Plates were incubated at 35°C for 24 h for the isolation of presumptive colonies. To confirm the presumptive *V. vulnificus vvha*+, *vcg* E, and *vcg* C, and *V. parahaemolyticus tlh*+, *tdh*+, *trh*+, and *orf*8, at least 15 blue-green and mauve well-grown colonies from each CHROMagar plates were selected and inoculated into APW tubes, incubated at 35°C for 18–24 h, and then subjected to DNA extraction, purification, and amplification. Presumptive strains that were confirmed with the direct PCR were scored as positive for the respective gene and stored in Trypticase soy agar (TSA; BIOXON Becton Dickinson S.A de C.V., Mexico) slants at −20°C. PCR assays were performed using specific primers (Sigma-Aldrich QUIMICA S.A. de C.V., Toluca, Mexico) for species and identification of pathogenic genes. Oligonucleotides targeting the *vvhA* (cytotoxin, cytolysin) and *tlh* (thermolabile hemolysin) genes were used for *V. vulnificus* and *V. parahaemolyticus*, respectively. Strains from the Collection of Aquatic Important Microorganisms (CAIM) of Centro de Investigaciones en Alimentación y Desarrollo A.C. Mazatlán, Sinaloa, México (www.ciad.mx/caim) were used as positive controls. DNA of strain CAIM 610 was used as positive control for (*vvha*) gene [28], and strains CAIM 1860 and 1859 for *vcgC* and *vcgE* genes, respectively [29]. DNA of strain CAIM 1772 was used as positive control for *V. parahaemolyticus* nontoxigenic (*tlh*) and toxigenic (*tdh*, *trh*) genes [30], and strain CAIM 1400 for *orf*8 gene [31]. A 100-bp ladder (100–3000 bp; Axygen) was used as a DNA size marker. Densities of *V. vulnificus* and *V. parahaemolyticus* strains were expressed by the most probable number (MPN) as *V. vulnificus* or *V. parahaemolyticus* MPN/g of oyster [32]. *V. vulnificus* and *V. parahaemolyticus* presumptive isolates were identified by biochemical characteristics using Kligler iron agar slants (KIA), lysine iron agar (LIA), motility-indole-ornithine medium (MIO), Moeller decarboxylase broth media, and arginine dihydrolase test. All agar media were BD Bioxon (Becton Dickinson de México S.A. de C.V., México, México). The oxidase test (p-aminodimethylaniline; Becton Dickinson, NJ, USA) was performed on growth from presumptively positive isolates. Some *V. vulnificus* strains isolated from oysters collected at the *Central de Abasto* market in Mexico City were characterized for *vvhA* and *vcg* genotype, using PFGE, multilocus sequence typing (MLST), and *rtx*A1. Analyses included a comparison with *rtx*A1 reference sequences. Environmental *V. vulnificus* C genotype strains had high similarity to the virulent reference strain (CAIM 1860) [33].

2.3 Statistical analysis

2.3.1 Seasonal densities and survival

Most probable number (MPN) three-tube chart and formulas corresponding 95% confidence limits were used to identify MPN for each sample [32]. MPN values for *V. vulnificus* and *V. parahaemolyticus* counts were log-transformed to normalize the data and homoscedasticity requirements for appropriate analysis of variance. Significant differences in the seasonal distributions of $\log_{10}$ MPN/g *V. vulnificus* and *V. parahaemolyticus* densities were analyzed by analysis of variance ($P < 0.05$) and Tukey's test. All statistical analyses were carried out with XLSTAT > 2018 software (Addinsoft™) with the minimum level of significance set at $P < 0.05$. Nondetectable values of *V. vulnificus* and *V. parahaemolyticus* counts (<0.30 MPN/g) were replaced by half of the detection limit in oysters for statistical purposes.

To evaluate the effect of the transportation time on the growth ($\log_{10}$ NMP/g) of *V. vulnificus* and *V. parahaemolyticus*, a modified Gompertz model was fitted to the experimental data obtained at the ambient temperatures of 20.1, 25.6, and 24.4°C [34] for 22 h during supply-chain transportation in windy, dry, and rainy seasons, respectively, from MLS to Mexico City. Lag time and specific growth rate of strains were determined using Statistica 7.0 (Statistica Software, Palo Alto, CA, USA). This model has been used to describe *V. parahaemolyticus* growth (Eq. (1)) [35]:

$$Y_0 = N_0 + A\, x\, exp\; [-exp(2.718 \frac{\mu_{max}}{A})\; x\; (\lambda - t) + 1] \qquad (1)$$

where Yt is the log counts (CFU/g) at time t; N_0 is the initial level of bacteria ($\log_{10}$ CFU/g); $A = \log_{10}(N_{max}/N_0)$, where N_{max} represents the growth from the inoculum to stationary phase; and the parameters exp, μ_{max}, and λ represent *e* constant, maximum specific growth rate (h^{-1}), and lag time of the strain growth (h), respectively. The effect of temperature on *V. vulnificus* and *V. parahaemolyticus* growth was calculated with Eq. (2):

$$G = \ln 2/\mu_{max} \qquad (2)$$

where G is the generation time (h) at 20.1, 25.6, and 24.4°C and μ_{max} is maximum specific growth rate (h^{-1}).

Goodness of fit of the modified model was evaluated using the coefficient of determination (R^2) and the standard deviation of the residuals (*Syx*), which were provided by Statistica software.

2.3.2 Risk assessment

The FDA/FAO/WHO v.2005 software in combination with Microsoft Excel was used to run the risk simulations using the model developed by the U.S. Food and Drug Administration and used by FAO/WHO to estimate the risk of illness associated to the consumption of raw oysters [37]. Results were expressed as number of cases per 100,000 servings (cocktails consumed). The consumption data considered *V. vulnificus* and *V. parahaemolyticus* levels in raw oysters at harvest and after transportation and serving size (a cocktail of 12 oysters).

3. Results and discussion

3.1 Seasonal densities and survival after transportation

As shown in **Table 1**, significant differences in *V. vulnificus vvha+* densities between seasons were observed, with higher ($P < 0.05$) mean levels during windy (0.720 $\log_{10}$ MPN/g) and the lowest in rainy (−0.523 $\log_{10}$ MPN/g) seasons. During windy season, the average water temperature in the MLS-Mata Grande bank was 25.6°C, nevertheless mean *V. vulnificus vvha+* densities decreased during rainy season when the average water temperature increased ($P > 0.05$) to 27.4°C. However, salinity was higher ($P < 0.05$) in windy (25.8‰) than in rainy (7.3‰) seasons (**Table 5**).

In contrast, *V. parahaemolyticus tlh+* density levels were high ($P > 0.05$) in rainy (0.713 $\log_{10}$ MPN/g) and low in windy (0.477 $\log_{10}$ MPN/g) seasons. After 22 h of supply-chain transportation, *V. vulnificus vvha+* and *V. parahaemolyticus tlh+* densities increased ($P < 0.05$) in all seasons probably due to the high ambient temperatures observed during transportation (20.1, 25.6, and 24.4°C). **Table 2** shows that no *V. vulnificus vvha+ vgc*E densities were detected at-harvest and remain

Seasons	*Vibrio vulnificus vvha+* ($\log_{10}$ MPN/g mean and range)		*V. parahaemolyticus tlh+* ($\log_{10}$ MPN/g mean and range)	
	At-harvest	**At-market**	**At-harvest**	**At-market**
Windy	$0.720 \pm 0.344^{a,x}$ (0.477–0.964)	$3.351 \pm 0.041^{a,y}$ (3.322–3.380)	$0.477 \pm 0.001^{a,x}$ (0.477–0.0.477)	$3.041 \pm 0.001^{a,y}$ (3.041–3.041)
Dry	$-0.483 \pm 0.056^{b,x}$ (−0.523 to −0.444)	$1.055 \pm 0.129^{b,y}$ (0.964–1.146)	$0.686 \pm 0.0.149^{a,x}$ (0.580–0.792)	$3.210 \pm 0.239^{a,y}$ (3.041–3.380)
Rainy	$-0.523 \pm 0.001^{c,x}$ (<−0.523)*	$3.351 \pm 0.041^{a,y}$ (3.322–3.380)	$0.713 \pm 0.221^{a,x}$ (0.556–0.869)	$3.380 \pm 0.001^{a,y}$ (3.380–3.380)

Means with different letter (a, b, c) are significantly different ($P < 0.05$) between seasons.
Means with different letter (x, y) are significantly different ($P < 0.05$) between hours of transportation within each season.
**<−0.523 = not detected.*

Table 1.
Seasonal variations of V. vulnificus vvha *and* V. parahaemolyticus tlh *densities ($\log_{10}$ MPN/g) in* Crassostrea virginica *samples from the MLS during 22-h supply-chain transportation in windy, dry, and rainy seasons, respectively, from MLS to Mexico City.*

Seasons	***Vibrio vulnificus vgc*E (log$_{10}$ MPN/g mean and range)**		***Vibrio vulnificus vgc*C (log$_{10}$ MPN/g mean and range)**	
	At-harvest	**At-market**	**At-harvest**	**At-market**
Windy	$-0.523 \pm 0.001^{a,x}$ (<−0.523)	$-0.483 \pm 0.056^{a,x}$ (−0.523 to −0.444)	$0.469 \pm 0.010^{a,x}$ (0.462–0. 477)	$0.781 \pm 0.005^{a,y}$ (0.778–0.785)
Dry	$-0.523 \pm 0.001^{a,x}$ (<−0.523)	$-0.523 \pm 0.001^{a,x}$ (<−0.523)	$-0.523 \pm 0.001^{b,x}$ (<−0.523)	$-0.523 \pm 0.001^{b,x}$ (<−0.523)
Rainy	$-0.523 \pm 0.001^{a,x}$ (<−0.523)	$-0.483 \pm 0.056^{a,x}$ (−0.523 to −0.444)	$-0.523 \pm 0.001^{b,x}$ (<−0.523)	$0.469 \pm 0.010^{c,y}$ (0.462–0. 477)
	V. parahaemolyticus tlh+/tdh+		***V. parahaemolyticus tlh+/tdh−/trh+***	
Windy	$-0.020 \pm 0.707^{a,x}$ (<−0.523–0.477)	$-0.523 \pm 0.001^{a,x}$ (<−0.523)	$-0.523 \pm 0.001^{a,x}$ (<−0.523)	$-0.523 \pm 0.001^{a,x}$ (<−0.523)
Dry	$-0.523 \pm 0.001^{a,x}$ (<−0.523)	$-0.523 \pm 0.001^{a,x}$ (<−0.523)	$-0.523 \pm 0.001^{a,x}$ (<−0.523)	$-0.523 \pm 0.001^{a,x}$ (<−0.523)
Rainy	$-0.523 \pm 0.001^{a,x}$ (<-0.523)*	$-0.484 \pm 0.056^{a,x}$ (−0.523 to −0.444)	$-0.523 \pm 0.001^{a,x}$ (<−0.523)	$-0.484 \pm 0.056^{a,x}$ (−0.523 to −0.444)

Means with different letter (a, b, c) are significantly different (P < 0.05) between seasons.
Means with different letter (x, y) are significantly different (P < 0.05) between hours of transportation within each season.
**<−0.523 = not detected.*

Table 2.
Seasonal variations of pathogenic V. vulnificus *(genotypes E and C) and* V. parahaemolyticus *(*tlh/tdh, tlh/trh*) densities (log$_{10}$ MPN/g) in* Crassostrea virginica *samples from the MLS during 22-h supply-chain transportation in windy, dry, and rainy seasons, respectively, from MLS to Mexico City.*

unculturable after 22-h transportation during dry season. A seasonal trend was observed, as higher ($P > 0.05$) *V. vulnificus vvha+ vgc*C density (0.469 log$_{10}$ MPN/g) in oysters harvested from Mata Grande bank was found during windy season, and no densities were detected during dry and rainy seasons. Similarly, *V. parahaemolyticus tdh+* density in oysters increased ($P > 0.05$) in windy season (−0.020 Log$_{10}$ MPN/g), but no densities were detected during dry and rainy seasons. In contrast, no *V. parahaemolyticus trh+* density was detected in all seasons. After 22 h of supply-chain transportation, a slight increase ($P > 0.05$) in *V. vulnificus vgc*E (−0.483 log$_{10}$ MPN/g) in windy and rainy seasons was observed. *V. vulnificus vgc*C density in oysters increased ($P < 0.05$) in windy (0.781 log$_{10}$ MPN/g) and rainy seasons (0.469 log$_{10}$ MPN/g) as well. An increase in densities of *V. parahaemolyticus tdh+* and *trh+* (−0.484 log$_{10}$ MPN/g) in oysters was observed in rainy season, probably due to the high ambient temperature observed (24.4°C) in rainy season. Our results were lower than those reported in oysters harvested from the U.S. Gulf of Mexico during dry season (3.36 log$_{10}$ MPN/g), which were higher than those detected during windy season (1.0 log$_{10}$ MPN/g) [37]. *V. vulnificus* proliferates in areas or during months where the water temperature exceeds 18°C as in MLS, and culturable concentrations of *V. vulnificus* are generally lower when water temperatures are cooler.

In other study, *V. vulnificus* was isolated from oyster samples collected from Pueblo Viejo Lagoon, Veracruz, and 27% (39/143) of the oyster samples were *vvha+*. Although positive samples were found during all seasons of a 1-year period, a seasonal fluctuation was observed. Isolation rates from oysters were significantly higher in June than in the period from November to February ($P < 0.0002$), indicating that water surface temperatures >24°C and salinity conditions >18‰ are more favorable for *V. vulnificus* [13]. In our study, we found higher ($P < 0.05$) *V. vulnificus vvha+* densities during windy (December to March) and dry seasons (April to July) when water temperature and salinity were 25.6°C/25.8‰ and 28.7°C/29.8‰, respectively. However, a decrease was observed during rainy season when water temperature and salinity were 27.4°C and 7.3‰, respectively. Thus, *V. vulnificus* colonization of oysters in MLS may be influenced by water parameters such as temperature or salinity as previously reported [38]. An important finding in our study was the isolation of pathogenic *V. vulnificus vgc*C strains. This is the first study to report the presence of *V. vulnificus vgc*C in oysters from the Mexican coastline of the Gulf of Mexico. It is unclear if levels of the two *V. vulnificus* genotypes are unique to certain environmental conditions. As with previous studies of total *V. vulnificus* levels, a significant negative correlation with salinity was observed for the *vgc*C strains from oysters ($r = -0.35$, $P = 0.008$) [39]. In our study, there was a significant increase in the population of *V. vulnificus vgc*C in oysters in winter season when MLS water salinity levels were high. These results seem to indicate that *V. vulnificus vgc*C strains have evolved to cope with the stresses associated with changing environment. The fact that oysters have *vgc*C strains as the dominant strain type further suggests the possibility that those oysters harboring larger densities of this genotype would likely to be more infective to humans as *V. vulnificus vgc*C type is more infectious [29].

Regarding *V. parahaemolyticus*, our results demonstrated that *V. parahaemolyticus tlh+* strains are present almost throughout the year as *V. parahaemolyticus* abundance in the Gulf of Mexico is almost constant because temperature is warmer (>11.6°C) [40]. The seasonal trends in *V. parahaemolyticus* densities observed in our study agree with previous studies since the seasonal cycle of the pathogen has been correlated with dry and rainy seasons in tropical waters, being salinity the major factor. *V. parahaemolyticus tlh+* density was detected at 3.26 $\log_{10}$ MPN/g in oysters (*Crassostrea brasiliana*) harvested from Sao Paulo, Brazil during the dry season when mean water temperature was 29°C and salinity 29‰ [41]. Our previous studies have shown that there is a seasonal variation in the survival and virulence of *V. parahaemolyticus*, probably caused by a response of gene expression to stress. *V. parahaemolyticus tlh+/tdh+* densities in oysters harvested from the MLS were observed during windy and dry seasons (0.97 and − 0.18 $\log_{10}$ MPN/g), respectively, and *V. parahaemolyticus tlh+/tdh−/trh+* (−0.37 $\log_{10}$ MPN/g) was only detected during dry season. Meanwhile, during rainy season only, −0.509 $\log_{10}$ MPN/g was identified [42]. The presence of pathogenic *V. parahaemolyticus* strains raises important health issues and may be indicative of high risk in usual consumers of oyster from Mandinga lagoon during windy season where the maximum densities are found. These data suggest that *V. vulnificus* and *V. parahaemolyticus* populations in oysters are controlled by different factors. Moreover, the oyster, as a living host, may have contributed to the variation in these pathogen densities because of fluctuations in physiology resulting from reproductive status, diet, and health [11]. Densities above Mexican limits (absence in 50 g of sample) [19] for *V. parahaemolyticus tlh+* and *V. vulnificus vvha+* were detected in oyster samples at-harvest and at-market. In Mexico, these pathogens are not currently included in the microbiological surveillance programs of shellfish from harvesting areas and they are also excluded from the Mexican communicable disease surveillance plans. Thus, the presence of pathogenic strains is a public health concern, as these strains are not covered by current regulations.

The values for the kinetic growth parameters and performance statistics of the modified Gompertz model for *V. parahaemolyticus* (*tlh+*, *tlh+/tdh+*, and *tlh+/tdh−/trh+*) and *V. vulnificus vvha+* and genotypes E and C, at ambient temperatures during 22 h transportation of oysters are shown in **Tables 3** and **4**, respectively. The average R^2 value of the model fitted to *V. parahaemolyticus* growth was 0.9999 for nonpathogenic *tlh+* and for pathogenic *tdh+* and *trh+* strains. Similarly, R^2 value of the model fitted to *V. vulnificus* growth was 0.9999 for *vvha+* and for *vcg*E and *vcgc* strains, indicating that this model was able to describe both pathogens growth. As shown in **Table 3**, the predicted lag time values of the nonpathogenic *tlh* strains were 4.2909, 4.3582, and 4.2484 h in windy, dry, and rainy seasons, respectively; meanwhile, the predicted lag time values for both pathogenic *tdh*+ and *trh*+ strains were 6.3439 during rainy season, indicating faster growth and better adaptation of the nonpathogenic strain to ambient temperatures during transportation.

V. parahaemolyticus	μ_{max} (h^{-1})	λ (h)	*A*	G (h)	R^2	S_{yx}
Windy						
tlh+	1.0242	4.2909 (257.5 min)	2.564	0.6767 (40.6 min)	0.9999	0.00067
Dry						
tlh+	1.0117	4.3582 (261.5 min)	2.520	0.6851 (41.1 min)	0.9999	0.00073
Rainy						
tlh+	1.0736	4.2484 (254.9 min)	2.670	0.6456 (38.7 min)	0.9999	0.00073
tlh+/*tdh*+	0.0096	6.3439 (380.6 min)	0.039	72.0207 (4321.2 min)	0.9999	0.00018
tlh+/*tdh*−/*trh*+	0.0096	6.3439 (380.6 min)	0.039	72.0207 (4321.2 min)	0.9999	0.00018

Table 3.
Parameter values using the modified Gompertz model for V. parahaemolyticus *(*tlh+, tlh+/tdh + *and* tlh+/tdh−/trh+*) growth in oysters transported for 22 h at 20.1, 25.6, and 24.4°C (windy, dry, and rainy seasons) from MLS to Mexico City.*

V. vulnificus	μ_{max} (h^{-1})	λ (h)	*A*	G (h)	R^2	S_{yx}
Windy						
vvha+	1.0592	4.2838 (257.0 min)	2.630	0.6544 (39.3 min)	0.9999	0.00757
*vcg*E	0.0098	6.3022 (378.1 min)	0.040	70.7009 (4242.1 min)	0.9999	0.00021
*vcg*C	0.0836	5.9274 (355.6 min)	0.31	8.2914 (497.5 min)	0.9999	0.00013
Dry						
vvha+	0.5280	4.5347 (272.0 min)	1.540	1.3126 (78.8 min)	0.9999	0.00621
Rainy						
vvha+	1.7885	4.3926 (263.6 min)	3.870	0.3876 (23.3 min)	0.9999	0.00730
*vcg*E	0.0098	6.3022 (378.1 min)	0.04	70.7009 (4242.1 min)	0.9999	0.00021
*vcg*C	0.3063	4.9150 (294.9 min)	0.990	2.2633 (135.8 min)	0.9999	0.00005

Table 4.
Parameter values using the modified Gompertz model for V. vulnificus *(*vvha+, vcg*E, and* vcg*C) growth in oysters transported for 22 h at 20.1, 25.6, and 24.4°C (windy, dry, and rainy seasons) from MLS to Mexico City.*

Pathogenic strains were detected in oysters after 22 h of transportation only during rainy season. These results indicated that nonpathogenic *tlh*+ and pathogenic *tdh*+ and *trh*+ strains reached a maximum growth rate and the maximum density (6.670, 0.039, and 0.039 $\log_{10}$ MPN/g, respectively) after 22-h transportation at ambient temperature during rainy season. The values of lag time observed in this study were lower than those previously reported for nonpathogenic *tlh*+ (24.6 h) and pathogenic *trh*+ (38.7 h) strains of *V. parahaemolyticus* isolated from raw Korean oysters [35]. In the present study, the longer lag time of pathogenic strains may be due to the time required for colonization of the oyster tissue. It has been reported that *V. parahaemolyticus* colonized oyster tissues according to the change of time as it is digested by oysters [43]. The maximum specific growth rate (max) predicted for pathogenic *tdh*+ and *trh*+ strains (0.0096 h^{-1}) was lower than that for nonpathogenic *tlh*+ (1.0242, 1.0117, and 1.0736 h^{-1}) in windy, dry, and rainy seasons, respectively; generation times (*G*) of nonpathogenic (0.6767, 0.6851, 0.6456 h) in windy, dry, and rainy seasons, respectively, were shorter than that for pathogenic strains (72.0207 h). These results indicated that nonpathogenic *V. parahaemolyticus* strains reached a maximum growth rate faster by storage temperatures. However, both pathogenic and nonpathogenic *V. parahaemolyticus* grew during storage in rainy season, although it was not detected in at-harvest oysters. This finding suggests that the bacterium was most likely present in numbers below the limit of detection, or perhaps in a viable but nonculturable state. In addition, it has been also observed that *V. parahaemolyticus* multiplied rapidly in live oysters held at 26°C after harvest [20]. It has been reported that higher concentrations of *V. parahaemolyticus* are present in market oysters than in at-harvest oysters [44]. In our study, pathogenic *V. parahaemolyticus* strains had the ability to adapt and survive at 24.4°C during transportation in rainy season, prior to marketing. However, the growth characteristics of *V. parahaemolyticus* might vary by strain variation.

According to **Table 4**, the predicted lag time values of *V. vulnificus vvha*+ strains were 4.2838, 4.3547, and 4.3926 h in windy, dry, and rainy seasons, respectively. The predicted lag time values were 6.3022 for *vcg*E and 5.9274 and 4.9150 for *vcg*C during windy and rainy seasons, respectively, indicating faster growth and better adaptation to ambient temperatures during transportation of the *vvha*+ than *vcg*C strains. No *vcg*C and *vcg*E strains were detected in oysters after 22 h of transportation during dry season. *V. vulnificus vcg*E strains lag time values were similar to those of *V. parahaemolyticus tdh*+ and *trh*+ strains, but higher than those of *V. vulnificus vcg*C strains. The maximum specific growth rate (max) predicted for *vcg*E (0.0098 h^{-1}) and *vcg*C strains (0.0836 and 0.3063 h^{-1}) were lower than that for *vvha*+ (1.0592, 0.5280, and 1.7885 h^{-1}) in windy, dry, and rainy seasons, respectively. Generation times (*G*) of *vvha*+ (0.3876 h), *vcg*E (70.7009 h), and *vcg*C strains (2.2633 h) in rainy season were shorter than that observed in windy and dry seasons. These results indicated that *V. vulnificus vvha*+, *vcg*, and *vcg*C strains reached a maximum growth rate and the maximum density (3.870, 0.04, and 0.990 log10 MPN/g, respectively) after 22-h transportation at ambient temperature during rainy season. It has been reported a maximal growth rate of 0.175/h and a 1.3 log10 increase in *V. vulnificus* levels in oysters stored at 28°C [45]. Recently, a predictive growth model for *V. vulnificus* in Pacific oysters was developed [46], where growth rate and lag time of *V. vulnificus* in shucked oyster meat at 24°C were 0.0138 h^{-1} and 5.38 h, respectively. Overall, this growth rate is much lower than those observed in *V. vulnificus vvha*+ strains in our study. However, the lag time value is higher than our *V. vulnificus vvha*+ strains values. *V. vulnificus* and *V. parahaemolyticus* densities in shell oysters at the stage of distribution were greater than those observed in oysters at-harvest. Moreover, both *V. vulnificus* and *V. parahaemolyticus* grew during storage, although they were not detected at-harvest oysters. During transport and storage of raw oysters, adverse conditions (low oxygen levels, accumulation of waste, and feeding interruption) are able to disrupt a variety of cellular processes and can promote the development of more stress-resistant cells, modulating the fitness and virulence of bacterial pathogens.

Studies have suggested that pathogenic strains have low levels of detection compared with nonpathogenic strains and are more sensitive to dystrophic conditions, rapidly becoming nonculturable [47]. Furthermore, differences in regulated genes between strains may more likely due to be a response against environmental stressors. Harvest and transport techniques used in this study were typical for the commercial oyster industry in the MLS and Alvarado Lagoon zones.

Therefore, these bacteriological findings in the commercial handling portion of the study are representative of the industry in Veracruz state and throughout perhaps the entire Mexican Gulf Coast oyster industry. These results indicate that the safety of raw oysters for consumption depends upon their initial degree of contamination, mainly due to the quality of seawater from which they are extracted or cultured, and to postharvest storage conditions. Because temperature abuse during postharvest handling and storage may increase the risk of illness due to the consumption of oysters, it is very important to predict the risk of *V. vulnificus* and *V. parahaemolyticus* to consumers. The infectious dose of *V. vulnificus* for the high-risk group is 2 log CFU/g [6]; therefore, for the protection of consumers, careful storage and consumption guidelines for oysters at retail markets and restaurants must be emphasized.

3.2 Risk assessment

According to **Table 5**, the results indicate that the risk of consuming a typical meal of 12 raw oysters contaminated with *V. vulnificus* would be higher in dry and rainy seasons. *V. vulnificus* levels in oysters and the corresponding consumer risk at the vending site are strongly influenced by climate, especially water and air temperatures. The findings indicate that the risk of oyster consumption from Veracruz, Mexico is slightly lower than those estimated by WHO/FAO [48] for *V. vulnificus* predicted to be associated with month- and year-specific water temperatures in the Gulf of Mexico, which were 3.37×10^{-5} and 4.28×10^{-5} during dry and rainy seasons, respectively. However, the risk of oyster consumption during windy season (2.9×10^{-6}) was similar to that reported by WHO/FAO (1.26×10^{-6}).

It is important to point out that seasonal expansion of *V. vulnificus* illnesses associated with oysters harvested from the Gulf of Mexico corresponds with warmer water temperatures (>20°C). The evidence indicates that climate anomalies have already greatly expanded the risk for vibrio illnesses [49]. WHO/FAO [48] reported a risk assessment for primary septicemia cases associated with consumption of raw oysters from the Gulf Coast of USA with mean densities of 57,000 and 80 MPN/g during summer and winter harvest seasons, respectively. In this context, variation

Season	Air temperature (°C)	Water temperature (°C)	Salinity (‰)	Risk for at-risk population (cases per 100,000 servings; 95% confidence interval)
Windy	20.1	25.6	25.8	2.9×10^{-6} (2.0×10^{-6}–3.8×10^{-6})
Dry	25.6	28.7	29.8	4.7×10^{-6} (3.8×10^{-6}–5.8×10^{-6})
Rainy	24.4	27.4	7.3	4.3×10^{-6} (3.5×10^{-6}–5.4×10^{-6})

Table 5.
Estimated risk assessment to V. vulnificus *associated to consumption of raw oyster cocktail expended at-harvest at the MLS and at-market in Mexico City during windy, dry, and rainy seasons.*

in water and air temperatures and the characteristics and temperature of transportation and storage time have the effect of increasing the variation of *V. vulnificus* numbers at each point along the harvest-to-consumption continuum.

Table 6 summarized the results of risk to *V. parahaemolyticus*. Results indicated that the contamination rates of virulent *V. parahaemolyticus* (*tdh*+ and *trh*+) in raw oysters at-harvest and at-market, and the transportation temperatures significantly influence the probability of illness. The risk of recently harvested oysters during windy season in Veracruz-Boca del Río oyster bars and restaurants where oysters harvested at the MLS are sold was 1.1-fold higher than the mean risk of consuming oysters during the rainy season. These results indicate that the risk of raw oyster consumption in Veracruz, Mexico is lower than those of the U.S. which were 4.4×10^{-4} [50], similar to those reported in Australia (6×10^{-8}–6.1×10^{-6}), higher than those of Canada (7.5×10^{-10}–1.1×10^{-6}) and New Zealand (8.6×10^{-8}–3.2×10^{-7}), but lower to that in Japan during autumn (1.2×10^{-4}) [51] and Taiwan (8.56×10^{-5}) [52]. The estimated risk in our study is similar to that reported in Malaysia (1.76×10^{-6}) [53], but lower than the average risk associated with the consumption of raw oysters contaminated with pathogenic *V. parahaemolyticus* marketed at Sao Paulo, Brazil of 4.7×10^{-4}, 6.0×10^{-4}, 4.7×10^{-4}, and 3.1×10^{-4} for spring, summer, fall, and winter, respectively [36]. As the microbial risk assessment was conducted, several limitations were identified. The estimation did not include the growth model of *V. vulnificus* and *V. parahaemolyticus* during the time gap from markets to consumption.

However, the model's assumption can be referred for retail outlets that serve fresh raw oysters where there is minimal time for the growth of both pathogens to occur. There is a growing body of evidence to suggest that *V. vulnificus* and *V. parahaemolyticus* infections are increasing and tend to follow regional climatic trends, with outbreaks following episodes of unusually warm weather. Moreover, several epidemiological factors, such as growing consumption and international trade of seafood produce, may increase the incidence of these pathogens [12]. In Mexico, there is currently a lack of detailed surveillance information regarding *V. vulnificus* and *V. parahaemolyticus* infections, which probably disguises their real clinical burden. However, there have been some reports of outbreaks and deaths caused by consumption of oysters contaminated with these pathogens. Recently, a patient with hepatic cirrhosis and hepatic carcinoma

Season	*Vibrio parahaemolyticus* density (log_{10} NMP/g)	Pathogenic rate (%)		Risk for at-risk population (cases per-100,000 servings; 95% confidence interval)	
		tdh+	*trh*+	*tdh*+	*trh*+
Windy					
At-harvest	−0.020	10.0	ND	8×10^{-6} (6.4×10^{-7}–1.0×10^{-4})	ND
At-market	ND	ND	ND	ND	ND
Rainy					
At-harvest	ND	ND	ND	ND	ND
At-market	−0.484	0.2	0.2	7.8×10^{-7} (6.2×10^{-8}–9.9×10^{-6})	7.8×10^{-7} (6.2×10^{-8}–9.9×10^{-6})

Table 6.
Estimated risk assessment to V. parahaemolyticus *associated to consumption of raw oyster cocktail expended at-harvest at the MLS and at-market in Mexico City during windy, dry, and rainy seasons.*

suffered fulminant sepsis and necrohemorrhagic bullae secondary to a *V. vulnificus* infection. The patient had ingested oysters in Mexico City 36 h before [54]. Along Veracruz state in Mexican Gulf Coast, 18 *V. parahaemolyticus* infections were reported. Of 18 patients, 27.7% (5/18) consumed raw oysters at oyster bars and restaurants located in Boca del Río and Veracruz Port [55]. The information provided in this study is important for preventing public health problems as pathogenic genes such as *vcg*C, *tdh*+ and *trh*+ were detected. Moreover, the distribution and variation in numbers of virulent *V. vulnificus* and *V. parahaemolyticus* in oysters may need to be determined before harvest as these data should be valuable for assessment of the human health risk due to consumption of raw oysters which represents a significant threat to human health and seafood safety.

4. Conclusions

The evidence indicates that there are significant differences in *V. vulnificus vvha*+ densities between seasons, with higher mean levels during windy and the lowest in rainy seasons. In contrast, *V. parahaemolyticus tlh*+ density levels were high in rainy and low in windy seasons. After 22 h of supply-chain transportation, *V. vulnificus vvha*+ and *V. parahaemolyticus tlh*+ densities increased due to the high ambient temperatures observed during transportation in all seasons. Pathogenic *V. vulnificus vvha*+ *vgc*C and *V. parahaemolyticus tdh*+ densities in oysters increased in windy season as well. After 22 h of supply-chain transportation, *V. parahaemolyticus tdh*+ and *trh*+ densities increased in rainy season, and *V. vulnificus vgc*C density in oysters increased in windy and rainy seasons. This is the first study to report the presence of *V. vulnificus vgc*C in oysters from the Mexican coastline of the Gulf of Mexico.

Densities above Mexican limits for *V. parahaemolyticus tlh*+ and *V. vulnificus vvha*+ were detected in oyster samples at-harvest and at-market. The presence of pathogenic strains is a public health concern, as these strains are not covered by current regulations. After 22-h transportation at ambient temperature during rainy seasons, nonpathogenic *V. parahaemolyticus tlh*+ and pathogenic *tdh*+ and *trh*+ strains and *V. vulnificus vvha*+, *vcg*E, and *vcg*C strains reached a maximum growth rate and the maximum densities. The risk of consuming a typical meal of 12 raw oysters contaminated with *V. vulnificus* would be higher in dry and rainy seasons, and during windy season for *V. parahaemolyticus*. Although the risk of recently harvested oysters from MLS during the windy, dry, and rainy seasons in Veracruz-Boca del Río oyster bars and restaurants is low, results indicated that the contamination rates of virulent *V. vulnificus* and *V. parahaemolyticus* in raw oysters at-harvest and at-market and the transportation temperatures significantly influence the probability of illness.

Adjustments in industry practices and regulatory policy should be considered, especially for seafood that is consumed raw, such as oysters. The results of this study could help Mexican regulatory agencies to develop sanitary norms to protect the population against health risks caused by consumption of raw oysters contaminated with pathogenic strains, and oyster processors to implement control measures. To reduce the risk of *V. vulnificus* and *V. parahaemolyticus* infection from consuming raw oysters, more rigorous postharvest time-temperature controls and surveillance during transportation and marketing of raw oysters must be implemented for immediate detection of these pathogens, especially if oysters are exported to other countries. In this context, the public should be educated by the local government that foodborne illness must never be measured as a minor illness. If measures for mitigating *V. vulnificus* and *V. parahaemolyticus* could not lead to a reduction of predicted risk associated with these pathogens and the global climate scenario worsens, the predicted risk will increase.

Acknowledgements

This work was supported by Mexican National Council of Science and Technology CONACYT and the Mexican Ministry of Health project research grant [114024] under the technical responsibility of Dr. Leonardo Lizárraga-Partida. We thank Centro de Investigaciones en Alimentación y Desarrollo A.C. (CIAD) of Vibriomex Group for supplying the control strains of *V. vulnificus* and *V. parahaemolyticus* from their Collection of Aquatic Important Microorganisms (CAIM, www.ciad.mx/caim).

Conflict of interest

The authors have no conflict of interest to declare.

Author details

Violeta Pardío[1*], Irma Wong[2], Leonardo Lizárraga[3], Karla López[1], Argel Flores[1], Guadalupe Barrera[4], Francisco Alarcón[1] and Carlos Fernández[4]

1 Faculty of Veterinary Medicine and Zootechnics, University of Veracruz, Veracruz, Mexico

2 National Autonomous University of Mexico, Institute of Ocean Science and Limnology, Mexico City, Mexico

3 Center of Scientific Research and Superior Education of Ensenada Baja California, Marine Biotechnology, Ensenada, Baja California, Mexico

4 Biological and Health Sciences Division, Metropolitan Autonomous University, Mexico City, Mexico

*Address all correspondence to: vpardio@uv.mx

References

[1] DePaola A, Hopkins L, Peeler J, Wentz B, McPhearson R. Incidence of *Vibrio parahaemolyticus* in US coastal waters and oysters. Applied and Environmental Microbiology. 1990;**56**:2299-2302

[2] Dileep V, Kumar H, Kumar Y, Nishibuchi M, Karunasaga I. Application of polymerase chain reaction for detection of *Vibrio parahaemolyticus* associated with tropical seafoods and coastal environment. Letters of Applied Microbiology. 2003;**36**:423-427

[3] Wang D, Yu W, Chen D, Zhang D, Shi X. Enumeration of *Vibrio parahaemolyticus* in oyster tissues following artificial contamination and depuration. Letters in Applied Microbiology. 2010;**51**:104-108

[4] Song E, Lee S, Lim H, Kim J, Jang K, Choi S, et al. *Vibrio vulnificus vvhA* induces autophagy-related cell death through the lipid raft-dependent c-Src/NOX signaling pathway. Scientific Reports. 2016;**6**:27080

[5] Johnson C, Bowers J, Griffitt K, Molina V, Clostio R, Pei S, et al. Ecology of *Vibrio parahaemolyticus* and *Vibrio vulnificus* in the coastal and estuarine waters of Louisiana, Maryland, Mississippi, and Washington (United States). Applied and Environmental Microbiology. 2012;**78**:7249-7257

[6] Horseman M, Surani S. A comprehensive review of *Vibrio vulnificus*: An important cause of severe sepsis and skin and soft-tissue infection. International Journal of Infectious Diseases. 2011;**15**:e157-e166

[7] Igbinosa E, Okoh O. Emerging Vibrio species: An unending threat to public health in developing countries. Research in Microbiology. 2008;**159**:495-506

[8] Su Y, Liu C. *Vibrio parahaemolyticus*: A concern of seafood safety. Food Microbiology. 2007;**24**:549-558

[9] Nordstrom J, Vickery M, Blackstone G, Murray S, DePaola A. Development of a multiplex real-time PCR assay with an internal amplification control for the detection of total and pathogenic *Vibrio parahaemolyticus* bacteria in oysters. Applied and Environmental Microbiology. 2007;**73**:5840-5847

[10] Chao G, Jiao X, Zhou X, Yang Z, Huang J, Pan Z, et al. Serodiversity, pandemic O3:K6 clone, molecular typing and antibiotic susceptibility of foodborne and clinical *Vibrio parahaemolyticus* isolates in Jiangsu, China. Foodborne Pathogens and Disease. 2009;**6**:1021-1028

[11] Froelich B, Oliver J. The Interactions of *Vibrio vulnificus* and the oyster *Crassostrea virginica*. Microbial Ecology. 2013;**65**:807-816

[12] Baker-Austin C, Stockley L, Rangdale R, Martinez-Urtaza J. Environmental occurrence and clinical impact of *Vibrio vulnificus* and *Vibrio parahaemolyticus*: A European perspective. Environmental Microbiology. 2010;**2**:7-18. DOI: 10.1111/j.1758-2229.2009.00096.x

[13] Quiñones-Ramirez E, Natividad I, Betancourt-Rule M, Ramirez-Vives F, Vazquez-Salinas C. Putative virulence factors identified in *Vibrio vulnificus* strains isolated from oysters and seawater in Mexico. International Journal of Environmental Health Research. 2010;**20**:395-405. DOI: 10.1080/09603123.2010.491856

[14] Torres M, Fernández E. Incidence of *Vibrio parahaemolyticus* in raw fish, oysters, and shrimp. Revista Latinoamericana de Microbiología. 1993;**35**:267-272

[15] Charles-Hernández G, Cifuentes E, Rothenberg J. Environmental factors associated with the presence of *Vibrio parahaemolyticus* in sea products and the risk of food poisoning in communities bordering the Gulf of Mexico. Journal of Environmental Health Research. 2006;**2**:1-7

[16] López K, Pardío V, Lizárraga L, Williams J, Martínez D, Flores A, et al. Environmental parameters influence on the dynamics of total and pathogenic *Vibrio parahaemolyticus* densities in *Crassostrea virginica* harvested from Mexico's Gulf coast. Marine Pollution Bulletin. 2015;**91**:317-329

[17] [CONAPESCA] Comisión Nacional de Acuacultura y Pesca. Anuario Estadístico de Acuacultura y Pesca. 2017. Available from: http://www.conapesca.sagarpa.gob.mx/ [Accessed: 08-01-2018]

[18] [SSA] Secretaría de Salud, Gobierno de México. NORMA Oficial Mexicana NOM-242-SSA1-2009. Productos y servicios. Productos de la pesca frescos, refrigerados, congelados y procesados. Especificaciones sanitarias y métodos de prueba. 2009. Available from: http://portal.salud.gob.mx/ [Accessed: 08-08-2018]

[19] Kinsey T, Lydon K, Bowers J, Jones J. Effects of dry storage and resubmersion of oysters on total *Vibrio vulnificus* and total and pathogenic (*tdh+/trh+*) *Vibrio parahaemolyticus* levels. Journal of Food Protection. 2015;**78**:1574-1580. DOI: 10.4315/0362-028X.JFP-15-017

[20] Gooch J, Depaola A, Owers J, Marshall D. Growth and survival of *Vibrio paraemolyticus* in postharvest American oysters. Journal of Food Protection. 2002;**65**:970-974

[21] Cook D. Refrigeration of oyster shellstock: Conditions which minimize the outgrowth of *Vibrio vulnificus*. Journal of Food Protection. 1997;**60**:349-352

[22] [SIPESCA] Sistema de Información de Pesca y Acuacultura. 2018. Available from: https://sipesca.conapesca.gob.mx/loginfiel.php [Accessed: 08-08-2018]

[23] Contreras Espinosa F, Castañeda O. Las lagunas costeras y estuarios del Golfo de México: Hacia el establecimiento de índices ecológicos. In: Caso M, Pisanty I, Ezcurra E, editors. Diagnóstico Ambiental del Golfo de México, México. SEMARNAT/INE/INECOL/HARP-Institute; 2004. p. 628

[24] [SE] Secretaría de Economía, Gobierno de México. NMX-FF-001-SCFI-2011 Productos de la pesca-ostion en concha vivo, y pulpa envasada-refrigerada (litoral Atlántico)-especificaciones. 2011. Available from: http://www.economia.gob.mx [Accessed: 03-01-2018]

[25] [SSA] Secretaría de Salud, Gobierno de México. NOM-109-SSA1-1994. Procedimiento para la toma, manejo y transporte de muestras de alimentos para su análisis microbiológico. 1994. Available from: http://portal.salud.gob.mx/ [Accessed: 03-01-2018]

[26] Lizárraga-Partida M, Gómez-Gil B, Méndez-Gómez E, Wong-Chang I, Pardío-Sedas V, Cabanillas-Beltrán H. Molecular detection of *V. cholerae*, *V. parahaemolyticus* and *V. vulnificus* in oyster from Mexico by the vibriomex group. In: Proceedings of the International Conference on Vibrios in the Environment; 7-12 November 2010; Biloxi, Mississippi, USA

[27] Barrera-Escorcia G, Wong-Chang I, Fernandez-Rendon C, Vazquez-Botello A, Gomez-Gil B, Lizarraga-Partida M. Quantification of Vibrio species in oysters from the Gulf of Mexico with two procedures based on MPN and PCR. Environmental Monitoring

and Assessment. 2016;**188**:602. DOI: 10.1007/s10661-016-5620-9

[28] Brauns L, Hudson M, Oliver J. Use of polymerase chain-reaction in the detection of culturable and nonculturable *Vibrio vulnificus* cells. Applied and Environmental Microbiology. 1991;**57**:2651-2655

[29] Rosche T, Yano Y, Oliver J. A rapid and simple PCR analysis indicates there are two subgroups of *Vibrio vulnificus* which correlate with clinical or environmental isolation. Microbiology and Immunology. 2005;**49**:381-389

[30] Bej A, Patterson D, Brasher C, Vickery M, Jones D, Kaysner C. Detection of total and hemolysin producing *Vibrio parahaemolyticus* in shellfish using multiplex PCR amplification of *tlh*, *tdh* and *trh*. Journal of Microbiological Methods. 1999;**36**:215-225

[31] Myers M, Panicker G, Bej A. PCR detection of a newly emerged pandemic *Vibrio parahaemolyticus* O3:K6 pathogen in pure cultures and seeded waters from the Gulf of Mexico. Applied and Environmental Microbiology. 2003;**69**:2194-2200

[32] [USDA] United States Department of Agriculture. [LQAD] Laboratory Quality Assurance Division. Most Probable Number Procedure and Tables. MLG Appendix 2.03. United State. 2008

[33] Guerrero A, Gómez Gil B, Wong Chang I, Lizárraga Partida M. Genetic characterization of *Vibrio vulnificus* strains isolated from oyster samples in Mexico. International Journal of Environmental Health Research. 2015;**25**:614-627

[34] [CONAGUA] Comisión Nacional del Agua. 2018. Available from: https://smn.cna.gob.mx/tools/DATA/Climatolog%C3%ADa/Pron%C3%B3stico%20clim%C3%A1tico/Temperatura%20y%20Lluvia/TMED/2012.pdf [Accessed: 03-01-2018]

[35] Yoon K, Min K, Jung Y, Kwon K, Lee J, Oh S. A model of the effect of temperature on the growth of pathogenic and nonpathogenic *Vibrio parahaemolyticus* isolated from oysters in Korea. Food Microbiology. 2008;**25**:635-641

[36] Costa Sobrinho P, Destro M, Franco D, Landgraf M. A quantitative risk assessment model for *Vibrio parahaemolyticus* in raw oysters in Sao Paulo State, Brazil. International Journal of Food Microbiology. 2014;**180**:69-77

[37] Johnson C, Flowers A, Noriea N, Zimmerman A, Bowers J, DePaol A, et al. Relationships between environmental factors and pathogenic vibrios in the northern Gulf of Mexico. Applied and Environmental Microbiology. 2010;**76**:7076-7084

[38] Pfeffer C, Hite M, Oliver J. Ecology of *Vibrio vulnificus* in estuarine waters of eastern North Carolina. Applied and Environmental Microbiology. 2003;**69**:3526-3531

[39] Warner E, Oliver J. Population structures of two genotypes of *Vibrio vulnificus* in oysters (*Crassostrea virginica*) and seawater. Applied and Environmental Microbiology. 2008;**74**:80-85

[40] Cabrera-García E, Vázquez-Salinas C, Quiñones-Ramírez E. Serologic and molecular characterization of *Vibrio parahaemolyticus* strains isolated from seawater and fish products of the Gulf of México. Applied and Environmental Microbiology. 2004;**7**:6401-6406

[41] Costa Sobrinho P, Destro M, Franco D, Landgraf M. Correlation between environmental factors and prevalence of *Vibrio parahaemolyticus* in oysters harvested in the southern coastal área

of Sao Paulo State, Brazil. Applied and Environmental Microbiology. 2010;**76**:1290-1293. DOI: 10.1128/AEM.00861-09

[42] Flores A, Pardío VT, Lizárraga L, López K, Uscanga R, Flores R. Seasonal abundance of total and pathogenic *V. parahaemolyticus* isolated from American oyster harvested in the Mandinga Lagoon System, Veracruz, Mexico: Implication for food safety. Journal of Food Protection. 2014;**77**:1069-1077

[43] Cabello A, Espejo R, Romero J. Tracing *Vibrio parahaemolyticus* in oysters (*Tiostrea chilensis*) using a green fluorescent protein tag. Journal of Experimental Marine Biology and Ecology. 2005;**327**:57-166

[44] Ellison R, Malnati E, DePaola A, Bowers J, Rodrick G. Populations of *Vibrio parahaemolyticus* in retail oysters from Florida using two methods. Journal of Food Protection. 2001;**64**:682-686

[45] Cook D. Effect of time and temperature on multiplication of *Vibrio vulnificus* in postharvest Gulf coast shellstock oysters. Applied and Environmental Microbiology. 1994;**60**:3483-3484

[46] Kim Y, Lee S, Hwang I, Yoon K. Effect of temperature on growth of *Vibrio parahaemolyticus* and *Vibrio vulnificus* in flounder, salmon sashimi and oyster meat. International Journal of Environmental Research and Public Health. 2012;**9**:4662-4675. DOI: 10.3390/ijerph9124662

[47] Raghunath P. Roles of thermostable direct hemolysin (TDH) and TDH-related hemolysin (TRH) in *Vibrio parahaemolyticus*. Frontiers in Microbiology. 2015;**5**:1-4

[48] [WHO/FAO] World Health Organization, Food and Agriculture Organization of the United Nations. Risk assessment of *Vibrio vulnificus* in raw oysters. Microbiological Risk Assessments 8. 2005. Available from: http://www.fao.org/docrep/008/a0252e/a0252e00.htm [Accessed: 08-01-2018]

[49] Martínez-Urtaza J, Bowers J, Trinanes J, DePaola A. Climate anomalies and the increasing risk of *Vibrio parahaemolyticus* and *Vibrio vulnificus* illnesses. Food Research International. 2010;**43**:1780-1790. DOI: 10.1016/j.foodres.2010.04.001

[50] [U.S FDA] U.S. Food and Drug Administration. Quantitative risk assessment on the public health impact of pathogenic *Vibrio parahaemolyticus* in raw oysters. 2005. Available from: http://www.fda.gov/food/scienceresearch/researchareas/riskassessmentsafetyassessment/uc%m050421.htm [Accessed: 03-01-2018]

[51] [WHO/FAO] World Health Organization/Food and Agriculture Organization of the United Nations. Risk Assessment of *Vibrio parahaemolyticus* in Raw and Under-Cooked Seafood: Interpretative Summary and Technical Report. Microbiological Risk Assessment Series No. 16. Rome: FAO; 2011

[52] Huang Y, Hwang C, Huang L, Wu V, Hsiao H. The risk of *Vibrio parahaemolyticus* infections associated with consumption of raw oysters as affected by processing and distribution conditions in Taiwan. Food Control. 2018;**86**:101-109. DOI: DOI.org/10.1016/j.foodcont.2017.10.022

[53] New C, Kantilal H, Tan M, Nakaguchi Y, Nishibuchi M, Son R. Consumption of raw oysters: A risk factor for *Vibrio parahaemolyticus* infection. International Food Research Journal. 2014;**21**:2459-2472

[54] Cervera-Gaviria J, Armenta-Alvarez A, Mendez-Ceballos A, Bolaños-Aguilar M, Martinez-Hernandez L, Virgen-Cuevas M, et al. Choque séptico secundario an infección por *Vibrio vulnificus*. Reporte de un caso. In: Proceedings of the XLI National Congress on Infectology and Clinical Microbiology; 25-28 May 2016; Monterrey, Nuevo León, México. p. S99

[55] [INAI] Instituto Nacional de Transparencia y Acceso a la Información. Secretaría de Salud: Listado nominal de resultados. México: Veracruz; 2016. Available from: https://www.infomex.org.mx/gobiernofederal/home.action [Accessed: 03-01-2018]